D0705886

Aristotle's *Physics*

Masterworks of Discovery
Guided Studies of Great Texts in Science
Harvey M. Flaumenhaft, Series Editor

Aristotle's *Physics*

A Guided Study

Joe Sachs

Rutgers University Press
New Brunswick and London

Seventh paperback printing, 2011

Library of Congress Cataloging-in-Publication Data

Sachs, Joe.
Aristotle's physics ; a guided study / by Joe Sachs.
p. cm. -- (Masterworks of discovery)
Includes index.
"A new translation, with introduction, commentary, and an explanatory glossary."
ISBN 0-8135-2191-2 (cloth). --ISBN 0-8135-2192-0 (pbk.)
1. Aristotle. Physics. English. 2. Physics--Early works to 1800. I. Aristotle. Physics. English. II. Title. III. Series.
Q151.A8S23 1995
530--dc20 94-46477
CIP

British Cataloging-in-Publication information available

Copyright © 1995 by Joe Sachs
All rights reserved
Manufactured in the United States of America

Contents

Note: The titles of the sections of the *Physics* have been added by the translator.

Series Editor's Preface

We often take for granted the terms, the premises, and the methods that prevail in our time and place. We take for granted, as the starting points for our own thinking, the outcomes of a process of thinking by our predecessors.

What happens is something like this: Questions are asked, and answers are given. These answers in turn provoke new questions, with their own answers. The new questions are built from the answers that were given to the old questions, but the old questions are now no longer asked. Foundations get covered over by what is built upon them.

Progress thus can lead to a kind of forgetfulness, making us less thoughtful in some ways than the people whom we go beyond: hence this series of guidebooks. The purpose of the series is to foster the reading of classic texts in science, including mathematics, so that readers will become more thoughtful by attending to the thinking that is out of sight but still at work in the achievements it has generated.

To be thoughtful human beings—to be thoughtful about what it is that makes us human—we need to read the record of the thinking that has shaped the world around us and still shapes our minds as well. Scientific thinking is a fundamental part of this record, but a part that is read even less than the rest. It was not always so. Only recently has the prevalent division between "the humanities" and "science" come to be taken for granted. At one time, educated people read Euclid and Ptolemy along with Homer and Plato, whereas nowadays readers of Shakespeare and Rousseau rarely read Copernicus and Newton.

Often it is said that this is because books in science, unlike those in the humanities, simply become outdated: in science the past is held to be passé. But if science is essentially progressive, we can understand it only by seeing its progress *as* progress. This means that we ourselves must move through its progressive stages. We must think through the process of thought that has given us what we otherwise would thoughtlessly accept as given. By refusing to be the passive recipients of ready-made presuppositions and approaches, we can avoid becoming their prisoners. Only by actively taking part in discovery—only by engaging in its re-discovery ourselves—can we avoid both blind reaction against the scientific enterprise and blind submission to it.

When we combine the scientific quest for the roots of things with the humanistic endeavor to make the dead letter come alive in a thoughtful mind, the past becomes a living source of wisdom that prepares us for the future—a more solid source of wisdom than vague attempts at being "interdisciplinary," which all too often merely provide an excuse for avoiding the study of scientific thought itself. The love of wisdom in its wholeness requires exploration of the sources of the things we take for granted, including the thinking that has sorted out the various disciplines, making demarcations between fields as well as envisioning what is to be done within them.

To help non-specialists gain access to formative writings in ancient and modern science, *Masterworks of Discovery* has been developed. The volumes of this series are not books *about* thinkers and their thoughts. They are neither histories nor synopses that can take the place of the original works. The volumes are intended to provide guidance that will help non-specialists to read for themselves the thinkers' own expressions of their thoughts. The volumes are products of a scholarship that is characterized by accessibility rather than originality, so that each guidebook can be read on its own without recourse to surveys of the history of science, to accounts of the thinkers' lives and times, to the latest scientific textbooks, or even to other volumes in the *Masterworks of Discovery* series.

While addressed to an audience that includes scientists as well as scholars in the humanities, the volumes in this series are meant to be readable by any intelligent person who has been exposed to the rudiments of high school science and mathematics. Individual guidebooks present carefully chosen selections of original scientific texts that are fundamental and thought provoking. Parts of books are presented when they will provide the most direct access to the heart of the matters that they treat. What is tacit in the texts is made explicit for readers who would otherwise be bewildered, or sometimes even be unaware that they should be bewildered. The guidebooks provide a generous supply of overviews, outlines, and diagrams. Besides explaining terminology that has fallen out of use or has changed its meaning, they also explain difficulties in the translation of certain terms and sentences. They alert readers to easily overlooked turning points in complicated arguments. They offer suggestions that help to show what is plausible in premises that may seem completely implausible at first glance. Important alternatives that are not considered in the text, but are not explicitly rejected either, are pointed out when this will help the reader think about what the text does explicitly consider. In order to provoke more thought about what are now the accepted teachings in science, the guidebooks bring forward questions about conclusions in the text that otherwise might merely be taken as confirmation of what is now prevailing doctrine.

Readers of these guidebooks will be unlikely to succumb to notions that reduce science to nothing more than an up-to-date body of concepts and facts and that reduce the humanities to frills left over in the world of learning after scientists have done the solid work. By their study of classic texts in science, readers of these guidebooks will be taking part in continuing education at the highest level. The education of a human being requires learning about the process by which the human race obtains its education—and there is no better way to do this than to read the writings of those master students who have been master teachers of the human race. These are the masterworks of discovery.

It was not in some simple and direct confrontation with a naturally given world that our modern natural science arose. Rather, it arose in confrontation with a world of human learning where Aristotle had been held to the be the master of those who know, and the master's teaching had been handed down in Latin by way of the Schoolmen of medieval Europe. The modern masters of discovery had to engage in a struggle demanding no little exertion of patience, fortitude, and cunning in order to win acceptance for strange new anti-Aristotelian teachings about nature that have long since become second nature for us products of the modern schools that have perpetuated their success. Those whom second nature bedazzles and empower find it very hard to see a first one.

So although we cannot fully understand the foundation of modern natural science unless we try to understand the older teaching that it buried, difficulties beset the effort to recover the ancient thinking that has been long dismissed as antiquated. It cannot be easy, in any case, to understand that radical and comprehensive effort at communicating primary insight which constitutes Aristotle's teaching about nature; but even before approaching the difficulties that are inherent in the enterprise, we may encounter other ones: avoidable difficulties resulting from a Latin tradition of translation that can keep us from seeing vividly what Aristotle was trying to say in

Greek about the world of nature as he saw it. With that in mind, the translation in this guidebook tries to obviate some obstacles interposed by the traditional mode of translation, so that an English-speaking reader who is willing to expend some effort can get a clearer, fresher view of Aristotle's thinking.

The *Physics* as presented in this guidebook is not a product of scholastic tradition. It is Aristotle's original attempt to explore nature as primarily experienced. It is, moreover, a dialectical inquiry rather than a deductive science. The reading of the *Physics* here proposed, which takes the text as presenting dialectical exploration rather than as intending conclusive demonstration, enables its argument to be seen as a unity, so that seemingly contradictory statements made in different parts of the text become stages of one inquiry, and can even be understood to serve the coherence of the whole. In particular, such a reading can reveal how closely connected are, on the one hand, Aristotle's justification for using the distinction between form and material in the argument of Book I, and, on the other hand, his definition of motion in Book III, his analysis of continuity in Book VI, and his discovery of a motionless first mover in Book VIII.

This guidebook contains a translation of Aristotle's *Physics* in its entirety, along with introductory remarks, appended explanations of terminology, and interspersed comment which overviews each part of the argument and indicates its significance for the argument as a whole, often concisely comparing and contrasting the remarks of Aristotle and those of other thinkers ancient and modern. The text that comes to sight here is not an ancient collection of primitive errors containing some crude anticipations of the science that we have inherited from the seventeenth century. What comes to sight, rather, is the setting out of a profound and coherent way of looking at the world, a way distinct from the one that is presupposed by later science. Discovering Aristotle's teaching in its original vigor enables us to get a better understanding of the modern science that emerged in opposition to it; and as we cease assuming an inevitable progress in which the later worldview simply corrects the earlier, we come to appreciate some deep questions raised by the choice between the worldviews as ways of seeking knowledge.

Studying Joe Sachs's unconventional translation of the *Physics*, and pondering the accompanying explanations and comments that are the fruits of his conviction that we have much to learn from Aristotle, readers may disagree with this or that suggestion offered by their guide, but his guidance will help them to take Aristotle's *Physics* far more seriously than it has been taken by many people for a long time. A text that might seem to be at most an interesting historical curiosity can become, for readers of this guidebook, a masterwork of discovery.

Harvey Flaumenhaft, Series Editor
St. John's College in Annapolis—August, 1994

Aristotle's *Physics*

Introduction

Socrates: In those days, when people were not wise like you young people, they were content to listen to a tree or a rock in simple openness, just as long as it spoke the truth, but to you, perhaps, it makes a difference who is speaking and where he comes from, for you do not concentrate on this alone: is that the way things are, or not?
Phaedrus: You are right to rebuke me.

Philosophic Writing

The activity we call *science* is dependent upon and embedded within a prior activity known as *philosophy*. Any scientific understanding presupposes opinions about the way things are. Those fundamental opinions, which must be the foundations of any science, are the direct topics of reflection in thinking that is philosophic. Philosophy is a permanent human possibility, and it must have arisen in all places and times when anyone paused in the business of life to wonder about things, but it was among the ancient Greeks that it was named and described and began to be reflected in written texts. It was two thinkers who wrote during the fourth century B.C., Plato and Aristotle, who showed the world for all time the clearest examples of philosophic thinking. Plato's dialogues display the inescapable beginnings of philosophy in all questions that touch on how a human being should live, and they show that such questions must open up for examination all the comfortable assumptions we make about the world. Aristotle's writings trace an immense labor of the intellect, striving to push the power of thinking to its limits.

Reading Aristotle, to be sure, is not at all like reading Plato. The dialogues are beautiful in style, sensitive in the depiction of living and breathing people, and altogether polished works meant for the widest public. The writings of Aristotle that we possess as wholes are school texts that, with the possible exception of the *Nicomachean Ethics,* seem never to have been meant for publication. The title that we have with the *Physics* describes it as a "course of listening." The likeliest conjecture is that these works originated as oral discourses by Aristotle, written down by students, corrected by Aristotle, and eventually assembled into longer connected arguments. They presuppose acquaintance with arguments that are referred to without being made (such as the "third man") and with examples that are never spelled out (such as the incommensurability of the diagonal). They are demanding texts to follow, and they are less interested in beauty of composition than in exactness of statement. But in the most important respect, the writings of Plato and Aristotle are more like each other than either is like anything else. Both authors knew how to breathe philosophic life into dead words on a page.

In Plato's dialogues, it is the figure of Socrates, always questioning, always disclaiming knowledge, always pointing to what is not yet understood, who keeps the tension of live thinking present. Despite the efforts of misguided commentators, one need only read any dialogue to see that there is no dogma there to be carried off, but only work to be done, work of thinking, into which Plato draws us. It may appear that Aristotle rejects this Platonic path, giving his thought the closure of answers and doctrine, turning philosophy into "science," but this is a distortion produced by transmission through a long tradition and by bad translations. The tradition speaks of physics, metaphysics, ethics, and so on as sciences in the sense of conclusions deduced from first principles, but the books written by Aristotle that bear those names contain no such "sciences." What they all contain is dialectical reasoning, argument that does not start with the highest knowledge in hand, but goes in quest of it, beginning with whatever opinions seem worth examining. Exactly like Plato's dialogues, Aristotle's writings lead the reader on from untested opinions toward more reliable ones. Unlike the author of the dialogues, Aristotle records his best efforts to get beyond trial and error to trustworthy conclusions. What keeps those conclusions from becoming items of dogma? The available translations hide the fact, but Aristotle devises a philosophic vocabulary that is incapable of dogmatic use.

This claim will come as a surprise to anyone familiar with the lore of substances and accidents, categories, essences, per se individuals, and so forth, but if Aristotle were somehow to reappear among us he would be even more surprised to find such a thicket of impenetrable verbiage attributed to him. Aristotle made his students work hard, but he gave them materials they could work *with*, words and phrases taken from the simplest contents of everyday speech, the kind of language that is richest in meaning and most firmly embedded in experience and imagination. The only trouble with ordinary speech, for the purposes of philosophy, is that it carries too much meaning; we are so accustomed to its use that it automatically carries along all sorts of assumptions about things, that we make without being aware of them. Aristotle's genius consists in putting together the most ordinary words in unaccustomed combinations. Since the combinations are jarring, our thinking always has to be at work, right now, afresh as we are reading, but since the words combined are so readily understood by everyone, our thinking always has something to work with. The meanings of the words in Aristotle's philosophic vocabulary are so straightforward and inescapable that two results are assured: we will be thinking about *something,* and not stringing together empty formulas, and we will be reliably in communication with Aristotle, thinking about the very things he intended.

We need to illustrate both the sort of thing that Aristotle wrote and the way the translations we have destroy its effect. Consider the word essence. It is an English word, and we all know more or less how to use it. Perfumes have essences, beef stock can be boiled down to its essence, and the most important part of anything can be called its essence. It seems to have some connection with necessity, since we occasionally dismiss something as not essential. By the testimony of usage, that is just about the whole of the word's meaning. Essense is a relatively vague English word. If we know Latin, the word begins to have some resonance, but none of that has crossed over into English. So what do we do when we find a translation of Aristotle full of the word *essence?* We have to turn to expert help. Ordinary dictionaries will probably not be sufficient; we will need philosophic dictionaries, commentaries on Aristole, textbooks on philosophy, or trained lecturers who possess the appropriate degrees. In short, we need to be initiated into a special club; it may make us feel superior to the ordinary run of human beings, and it will at least make us think that philosophy is not for people in general, but only for specialists. Medical doctors, for example, seek just those effects for their area of expert knowledge by never using an ordinary, understandable name for anything, but only a Latin derivative with many syllables. But did Aristotle want such a result? If not, writing in such a style can hardly be presented as a translation of Aristotle.

What did Aristotle write where the translators put the word *essence?* In some places he wrote "the what" something is, or "the being" of it. In most places he wrote "what something keeps on being in order to be at all," or "what it is for something to be." These phrases bring us to a stop, not because we cannot attach meaning to them, but because it takes some work to get hold of what they mean. Since Aristotle chose to write that way, is it not reasonable to assume that he would want us to do just that? In the poet Gerard Manley Hopkins's line "Though worlds of wanwood leafmeal lie," everything in his words is readily accessible even though the pieces are combined in unusual ways. We recognize this sort of word-play as a standard device of poetry, which works on us through the ear, the visual imagination, and our feelings. The poet makes us experience a fresh act of imagining and feeling, at his direction. (Think of all that would be lost if he had written "Notwithstanding the fact that an immeasurable acreage of deciduous forest manifests the state of affairs characteristic of its incipiently dormant condition.") Aristotle's phrases in the present example do something that is exactly analogous to the poet's word-play, but it is directed only at the intellect and understanding. Other words and phrases of his do carry imaginative content, but are subordinated to the intellect and understanding. Aristotle is not a poet, but a philosophic writer, one who, like a poet, loosened and recombined the most vivid parts of ordinary speech to make the reader see and think afresh. Many philosophers have written books, but few have worked as carefully and deliberately to make the word be suited to the philosophic deed.

Translation and Tradition

A long stretch of centuries stands between Aristotle and us. The usual translations of his writings stand as the end-product of all the history that befell them in those centuries. For about five centuries up to 1600 they were the source of the dominant teachings of the European universities; for about four centuries since then they have been reviled as the source of a rigid and empty dogmatism that stifled any genuine pursuit of knowledge. One has to be very learned indeed to uncover all that history, but fortunately for those of us who are interested only in understanding the writings themselves, no such historical background is of any use. In fact, it takes us far away from anything Aristotle wrote or meant. By chance, when Aristotle's books dominated the centers of European learning, the common language of higher learning was Latin. When in turn later thinkers rebelled against the tyranny of the established schools, it was a Latinized version of Aristotle that they attacked. They wrote in the various modern European languages, but the words and phrases of Aristotle that they argued with and about came into those languages with the smallest possible departures from the Latin.

Thomas Hobbes, for example, writing in 1651 (in the next to last chapter of the last part of *Leviathan),* makes a common complaint in a memorable way:

> I beleeve that scarce any thing can be more absurdly said in naturall Philosophy, than that which is now called Aristoteles Metaphysiques. . . . And since the Authority of Aristotle is onely current [in the universities], that study is not properly Philosophy, (the nature whereof dependeth not on Authors,) but Aristotelity. . . . To know now upon what grounds they say there be Essences Abstract, or Substantiall Formes, wee are to consider what those words do properly signifie. . . . But what then would become of these Terms, of Entity, Essence, Essentiall, essentiality, that . . . are . . . no Names of Things. . . . [T]his doctrine of Separated Essences, built on the Vain Philosophy of Aristotle, would fright [men] . . . with empty names as men fright Birds from the Corn with an empty doublet, a hat, and a crooked stick.

The usual translations of Aristotle are concerned most of all with preserving a continuity of tradition back through these early modern critics of Aristotle. Richard McKeon, in a note to a philosophic glossary, defends this practice:

> The tendency recently in translations from Greek and Latin philosophers, has been to seek out Anglo-Saxon terms, and to avoid Latin derivatives. Words as clear and as definitely fixed in a long tradition of usage as privation, accident, and even substance, have been replaced by barbarous compound terms, which awaken no echo in the mind of one familiar with the tradition, and afford no entrance into the tradition to one unfamiliar with it. In the translations above an attempt has been made to return to the terminology of the . . . English philosophers of the seventeenth century. Most of the Latin derivatives which

> are used . . . have justification in the works of Hobbes, Kenelm Digby, Cudworth, Culverwell, even Bacon, and scores of writers contemporary with them. . . . [T]he mass of commentary on Aristotle will be rendered more difficult, if not impossible, of understanding if the terms of the discussion are changed arbitrarily after two thousand years. (*Selections from Medieval Philosophers,* Vol. 2, New York: Scribner's, 1930, pp. 422–23)

The tendency deplored by McKeon has not made its way into any translations of the writings of Aristotle known to this writer. There was some hope of it when Hippocrates Apostle, announcing a new series of translations, included the following among its principles: "The terms should be familiar, that is, commonly used and with their usual meanings. If such terms are available, the use of strange terms, whether in English or in some other language, adds nothing scientific to the translation but unnecessarily strains the reader's thought and often clouds or misleads it" *(Aristotle's Metaphysics,* Bloomington: Indiana University Press, 1966, p. x). This sentiment is worth endorsing, but Apostle respects it only to the minor extent of avoiding such pretentious phrases as *ceteris paribus* (Latin for "other things being equal"), and nothing in his translations would disturb McKeon. But if Apostle's general claim is correct, and if in addition Aristotle himself never used technical jargon, then surely to use such language to translate him is to confuse Aristotle's writings with a tradition that adapted them to purposes that were not his. And if that Latin tradition distorted Aristotle's meaning and was untrue to his philosophic spirit, until all that remained was the straw man so easily ridiculed by Hobbes and every other lively thinker of his time, then to insist on keeping Aristotle within the confines of that caricature is perverse.

No text can be translated from one language to another with complete accuracy, and an author who takes liberties with common usage in his or her own language is especially difficult to translate. But in this case there is one simple rule that is easy to follow and always tends in a good direction: avoid all the conventional technical words that have been routinely used for Aristotle's central vocabulary. In fact, virtually all those words are poor translations of the Greek they mean to stand for. The word *privation,* for instance, is not found in this translation for the simple reason that its meaning cannot be expected to be known to all educated readers of

English. Commentaries on Aristotle use the word extensively, but if the Greek word it refers to has been adequately translated in the first place, you will not need commentaries to tell you what it means. Here that Greek word is translated sometimes as *deprivation,* sometimes as *lack,* according as one or the other fits more comfortably into its context. What matters is not whether Latin or Anglo-Saxon derivatives are used, but whether an understandable English word translates an understandable Greek one. *Accident* is a perfectly good English word, but not in the sense in which it appears in commentaries on Aristotle; the Greek word it replaces has a broad sense that corresponds to our word *attribute,* and a narrower one that can be conveyed by the phrase "incidental attribute." In this case again, Latin derivatives are available that carry clear and appropriate meanings in English, since one does not need to know any Latin to ferret them out. It is true that one sense of *ad-cadere* could have given rise to the meanings we attach to the words *incidental* and *attribute,* but it did not in fact transmit that meaning to its English derivatives. There is some pedantic pleasure in pointing out those connections, but to use the word *accident* in that sense is to write a forced Latin masquerading as English, guaranteed to confuse the non-specialist reader, where Aristotle used the simplest possible language in a way that keeps the focus off the words and on the things meant by them.

But to undo the mischief caused by McKeon's third example, substance, stronger medicine is required. Joseph Owens records the way this word became established in the tradition *(The Doctrine of Being in the Aristotelian 'Metaphysics,'* Toronto: Pontifical Institute of Mediaeval Studies, 1951, pp. 140–43). It is a comedy of errors in which Christian scruples were imposed upon a non-Biblical theology, and a disagreement with Aristotle was read back into his words as a translation of them. This translator ignores the contortions of the tradition and without apology uses the barbarous Anglo-Saxon compound *thinghood.* Though it occurs only a few times in the *Physics,* it is a central notion in the *Metaphysics* and in all Aristotle's thought, and the word *substance* does nothing but obscure its meaning. Lively arguments about substance go on today in the secondary literature, but a choice must be made, and the primary texts of Aristotle are clearer, richer, deeper, easier to absorb, and more worth pursuing than the commentaries on them. As it stands in the usual translations, the word substance is little more than an unkown *x* for which meaning has to be deduced by a kind of algebra, while Aristotle shows (*Metaphysics* 1028b, 2–7) that just asking what the thinghood of things consists in, and what is responsible for it, unlocks the highest inquiry of which philosophy is capable. For the promise of such a return, it is worth risking a little barbarity. The barbarism of a word like thinghood is just the fact that it falls far outside common usage in our language, and not in a direction that needs any historical or technical special knowledge to capture it, but in one that invites the same flexibility that poets ask of us. We cannot read such a word passively but must take responsibility for its meaning. This in itself, in moderation and in well-judged places, is something good and is an imitation of what Aristotle does with Greek.

It has already been remarked that the present translation does not always use the same English word for the same Greek one. This is partly because no English word ever has the same full range of meaning as any Greek word, so that such a range has to be conveyed, or unwanted connotations suppressed, by the use of a variety of near-synonyms. It is partly because a Greek word may have two or more distinct uses that differ according to context; in this way, the word for thinghood will often be translated as "an independent thing." It is also partly because Aristotle always paid attention to the fact that important words are meant in more than one way. For him that was not a fault of language, but one of the ways in which it is truthful. A word often has a primary meaning and a variety of derivative ones, as a reflection of causal relations in the world. A diet can be healthy only because, in a different and more governing sense, an animal can be healthy, and there can be a medical knife only because there is a medical skill. This array of difference within sameness usually cannot be lifted over from Greek to English, and has to be gotten at indirectly. For example, in the *Physics* every kind of change is spoken of as a motion, though the word for motion is gradually and successively limited until it refers strictly only to change of place. This progression determines the main structure of the inquiry, but in English the path would not be as clearly indicated by transitions of meaning within a single word. Finally, some words have many translations that are equally good in their different ways. In such cases, this translation rejoices in variety; this again is an imitation of Aristotle's general practice. Where the traditional translations are marked by rigid, formulaic repetitions, Aristotle loves to combine overlapping meanings, or separate intertwined meanings, to point to things his language had no precise word for. It is the living, natural, flexible character of thinking that breathes through Aristotle's use of language, and not the artificial, machine-like fixity one finds in the translations.

This last point should be taken not as a promise of smooth English but rather of the reverse. Idiomatic expressions and familiar ways of putting words together conceal unthinking assumptions of just the kind that philosophy tries to get beyond. The reader will need a willingness to follow sentences to places where meaning would be lost if it were forced into well-worn grooves, and will need to follow trains of thought that would not be the same if they did not preserve Aristotle's own ways of connecting them. As far as possible, this translation follows the syntax of Aristotle's text. Montgomery Furth followed this same procedure in a translation of part of the *Metaphysics,* and he apologizes that the result is neither English nor Greek, but Eek *(Aristotle, Metaphysics, Books VII–X,* Indianapolis: Hackett, 1985, p. vi). Furth translated this way in order to follow Aristotle's logic faithfully, but he retained all the usual Latinized vocabulary of Ross's Oxford translation, so that the resulting language might better be called Leek. The present translation goes farther, in vocabulary and syntax, beyond the Latin and toward the Greek, and could be called Gringlish, but for this as well it comes before you without apology. Furth's translation violates English sensibilities for the special purposes of graduate students and professional scholars; this translation violates them for the common human purposes of joining Aristotle in thinking that breaks through the habitual and into the philosophic.

A Philosophic Physics?

It may seem odd to combine philosophic aims with the topic of physics. It may seem that Aristotle had to speculate philosophically about the natural world because he did not have the benefit of the secure knowledge we have about it. In the current secondary literature one sees that at least some scholars think they might learn something about thinking from *De Anima* or about being from the *Metaphysics,* but articles on the *Physics* seem at most to pat Aristotle on the head for having come to some conclusion not utterly in conflict with present-day doctrines. This kind of smugness is a predictable result of the way the sciences have been taught to us. Conjectures and assumptions, because they have been part of authoritative opinion for a few centuries, are presented to us as stories, or as facts, without recourse to evidence or argument. Particular doctrines, even when they stand on theoretical structures as complex and fragile as a house of cards, or even when they presuppose a picture of things that is flatly in contradiction with itself, tend to be prefaced with the words "we know. . . ." All the rhetoric that surrounds the physics of our time tells us that philosophic inquiry need not enter its territory, that here the philosophizing is over and done, that the best minds agree about everything, and that, in any case, non-experts cannot hope to understand enough to assess the evidence. Strangely, the physics of the twentieth century is surrounded by the same air of dogmatic authority as was the school Aristotelianism of the sixteenth century.

But there are two kinds of support for the present-day physics that seem to lift it above dogmatism. One is a long history of experiment and successful technology, and the other is the greatest possible reliance on mathematics. These are both authorities that cannot be swayed by human preferences, and cannot lie. Their testimony, however, can be misunderstood, and can be incorporated into a picture of the world that fails in other ways. But even if the current physics contains nothing untrue, one might wish to understand it down to its roots, to unearth the fundamental claims about things on which it rests, which have been lost sight of in the onrush of theoretical and practical progress. To do this one has to stand back from it, to see its founding claims as alternatives to other ways of looking at the world, chosen for reasons. The earliest advocates of the "new physics" did just that, and the alternatives they rejected all stem from Aristotle's *Physics.* Martin Heidegger said that "Aristotle's *Physics* is the hidden, and therefore never adequately studied, foundational book of Western philosophy" ("On the Being and Conception of *Phusis* in Aristotle's *Physics* B, 1," in *Man and World,* 9, No. 3, (1976), p. 224). The physics of our times is inescapably philosophic, if only in the original choices, preserved in it, to follow certain paths of thought to the exclusion of others. To see that physics adequately and whole, we too need to be philosophic, to lift our gaze to a level at which it can be seen to be one possibility among many. Only then can we rationally and responsibly decide whether to adopt its opinions as our own.

But there is a second respect in which twentieth-century physics has opened its doors to philosophy and will not be able to close them. The physics that came of age in the seventeenth century, and seemed to have answered all the large questions by the nineteenth, is limping toward the end of the twentieth century in some confusion. Mathematics and technology have coped with all the crises of this century, but the picture of the world that underlay them has fallen apart. It was demonstrated conclusively that light is a wave, except when it shines on anything; then it arrives as particles. It is shown with equal certainty that the electron is a particle, except when it bounces off a crystal surface; then it must be a wave, interacting with the surface everywhere at once. Just when atomic physics seemed ready to uncover the details of the truth underlying all appearances, it began to undercut all its own assumptions. A wave-mechanics that held out an initial promise of reducing all appearances of particles to the behavior of waves failed to do so, and degenerated into a computational device for predicting probabilities. The most far-seeing physicists of the century have shown that particles and waves are equally necessary, mutually incompatible aspects of every atomic event, and that physics, at what was supposed to be the ultimate explanatory level, must abandon its claim to objectivity. The physicist is always describing, in part, his or her own decisions to interfere with things in one way rather than another; this brings along, as a causally necessary conclusion, the collapse of the belief in causal determinism. When Hobbes laughed at Aristotle, he was certain that he knew what a *body* is. Today all bets are off.

But some physicists have been unwilling to give up their dogmatic habits without a fight. Even Einstein, after he had taught the world to give up the rigid Newtonian ideas of time, space, and mass, was unable to suspend his unquestioned assumption that bodies have sharply defined places and cannot interact except by contact or by radiation. Niels Bohr and Werner Heisenberg had announced the most radical of revolutions, requiring physicists to ask what knowledge is, and no longer to answer by pointing to what they do. In a famous 1935 collaboration ("Can Quantum-Mechanical Description of Reality Be Considered Complete?" *Physical Review* 47, pp. 777–80), Einstein tried to hold off this final revolution, saying in effect "I know enough about the fundamental structure of the world to be certain that some things cannot happen." But experiments have revealed that those very things do happen, that the state of one particle is provably dependent on whether a measurement is performed on a distant second particle, from which no signal could have radiated.

A new and opposite tactic permits some physicists to embrace this or any other seeming impossibility without admitting the need for any philosophic re-thinking of the way things are. Listen to these words of Richard Feynman: "We always have had a great deal of difficulty in understanding the world view that quantum physics represents. At least I do, because I'm an old enough man that I haven't got to the point that this stuff is obvious to me. . . . You know how it always is, every new idea, it takes a generation or two until it becomes obvious that there's no real problem" (quoted by N. David Mermin in *The Great Ideas Today,* 1988, p. 52). So if the discoveries of quantum physics make you feel an urgent need to re-examine the presuppositions of physics, just repress that feeling for a generation or two, and it will go away.

Perhaps the strongest motive for the resistance to opening physics to philosophic examination is the plain fact that there is no need for physics to do anything differently. Whatever happens can be described mathematically, and new discoveries are readily incorporated into some mathematical scheme, and then predicted. Technology advances no less rapidly in areas in which the explanatory ground has been cut from under our feet than in those in which its workings are intelligible. But the new physics arose out of a desire to know, and it has undeniably become a highly questionable kind of knowing. Indeed, the very fact that its picture of the world can collapse while its mathematical description and practical applications are left intact is a powerful stimulus to wonder. While wishing physics well in all its uses, some of us may simply want to understand what it is and what it is not. But we cannot see how the various strands of physics have separated without understanding what it was to begin with, so again we are thrown back to the choices by which it came into being, and thus in turn to the picture of the world that it rejected. From this standpoint, though, such investigation is more than a quest to uncover something past and superseded. It entails the risk of being convinced that the original decisions of the seventeenth-century physicists are not all worthy of our own acceptance. It is possible that parts of Aristotle's understanding of the world might help heal our own dilemmas and confusions.

The Things That Are

Where should an understanding of the things around us begin? It might seem that there are plain facts that could serve as uncontroversial starting points. What are some of the plainest ones? The stars circle us at night, the sun by day. Rocks fall to earth, but flames leap toward the sky. Bodies that are thrown or pushed slow down continually until they stop moving. Animals and plants belong to distinct kinds, which are preserved from generation to generation. The visible whole is a sphere, with the earth motionless at its center. These are facts of experience, so obvious that the only way to be unaware of them is by not paying attention. If you disbelieve any of them it is not because of observation, but because you were persuaded not to trust your senses. No physics begins by looking at the things it studies; those things must always be assigned to some larger context in which they can be interpreted. Aristotle states this in the first sentence of his *Physics* by saying that we do not know anything until we know its causes. Nothing stands on its own, without connections, and no event happens in isolation; there must be some comprehensive order of things in which things are what they are and do what they do. Physics seeks to understand only a part of this whole, but it cannot begin to do so without some picture of the whole.

But it has been noted earlier that none of Aristotle's inquiries begins with the knowledge that most governs the things it studies. We never start where the truth of things starts, but must find our way there. But although we must be ready to modify our views as the inquiry proceeds, we cannot dispense with some preliminary picture of things. What is Aristotle's preliminary picture of the whole of things? It is one that permits the plainest facts of experience to be just the way they appear to us. We live at the center of a spherical cosmos as one species of living thing among many, in a world in which some motions are natural and some forced, but all require causes actively at work and cease when those causes cease to act. The natural motions are those by which animals and plants live and renew their kinds, the stars circle in unchanging orbits, and the parts of the cosmos—earth, water, air, and fire—are transformed into one another by heat and cold, move to their proper places up or down, and maintain an ever-renewing equilibrium. This picture is confirmed and fleshed out by Aristotle's inquiries in writings other than the *Physics,* but since Aristotle never writes "scientifically," that is, deductively, there is no necessary or right order in which they should be read. All those inquiries stand in a mutual relation of enriching and casting light upon one another, and the *Physics* is in an especially close relation with the *Metaphysics.*

The *Physics* assumes not only a picture of the whole, but also a comprehensive understanding of the way things are. In the *Metaphysics,* this comprehensive understanding is not assumed but is arrived at by argument, through the sustained pursuit of the question, What is being? Since being is meant in many ways, Aristotle looks for the primary sense of it, being as such or in its own right, on which the other kinds of being are dependent. That primary sense of being is first identified as thinghood and then discovered to be the sort of being that belongs only to animals, plants, and the cosmos as a whole. For these pre-eminent beings, being is being-at-work, since each of them is a whole that maintains itself by its own activity. For any other sort of being, what it is for it to be is not only something less than that, but it is in every case dependent on and derived from those highest beings, as a quality, quantity, or action of one, a relation between two or more, a chance product of the interaction between two or more, or an artificial product deliberately made from materials borrowed from one or more of them. Life is not a strange by-product of things, but the source of things, and the nonliving side of nature has being in a way strictly analogous to life: as an organized whole that maintains itself by continual activity.

In the central books of the *Metaphysics,* Aristotle captures the heart of the meaning of being in a cluster of words and phrases that are the most powerful expressions of his thinking. The usual translations of them not only fall flat but miss the central point: that the thinghood (*ousia*) of a thing is what it keeps on being in order to be at all (*to ti ēn einai*), and must be a being-at-work (*energeia*) so that it may achieve and sustain its being-at-work-staying-itself (*entelecheia*). In the standard translations of those words and phrases, that rich and powerful thought turns into the following mush: the substance of a thing is its essence, and it must be an actuality, so that it may achieve and sustain its actuality.

Once Aristotle's central thinking has been grasped, one can see that the physics that emerged in the seventeenth century adopted, in the principle of inertia, an understanding of being exactly opposite to that of Aristotle. The primary beings are what they are passively, by being hard enough to resist all change, and they do nothing but bump and move off blindly in straight lines. The picture of the world assumed by this physics is of atoms in a void, so there can be no cosmos, but only infinite emptiness; no life, but only accidental rearrangements of matter; and no activity at all, except for motion in space. This is an ancient idea, which goes back long before Aristotle's time. Some centuries later Lucretius found it appealing as a doctrine that teaches us that, while there is little to hope for in life except freedom from pain, there is little to fear either, since a soul made of atoms will dissolve, but cannot suffer eternal torment. There are reasons of two other kinds that make this picture of things attractive to the new physics. First, it makes it unnecessary to look for causes. Just because everything is taken to be reducible to atoms and the void, every possible event is pre-explained. Mechanical necessity takes over as the only explanation of anything, so the labor of explanation is finished at one stroke. And second, this picture makes every attribute of anything and every possible event entirely describable by mathematics. The glory of the new physics is the power it gains from mathematics. The world that is present to the senses is set aside as "secondary," and the mathematical imagination takes over as our way of access to the true world behind the appearances. The only experience that is allowed to count is the controlled experiment, designed in the imagination, with a limited array of possible outcomes that are all interpreted in advance.

From its beginnings, mathematical physics moves from success to success, but almost from the beginning its mechanistic picture of things fails. Newton begins his *Principia* with the assumption that all bodies are inert, but in the course of it he shows that every body is the seat of a mysterious power of attraction. Is this simply a new discovery to be added to our picture of the world? Shall we say that there are atoms, void, and a force of gravitation? But the whole purpose of the new world-picture was to avoid occult qualities. And where do we put this strange force of attraction? There is no intelligible way that inert matter can be conceived as causing an urge in a distant body. Shall we say that the force resides in a field? A field of what? The *Principia* shows that the spaces through which the planets move are void of matter. How can a point in empty nothing be the bearer of a quantity of energy? This new discovery can be described mathematically, but it does not fit into the world-picture that led to it, and cannot be understood as something added to it. Something similar happens with light, which was discovered by Maxwell to be describable as an electromagnetic wave. But a wave is a material conception: a disturbance in a string, or a body of water, or some such carrier, moves from one place to another while the parts of the body stay where they were. So when it was shown that a light-bearing aether would need to have contradictory properties, electromagnetic radiation was left as a well-described wave motion taking place in nothing whatever.

In the twentieth century the mechanist picture underlying mathematical physics has broken down even more radically, in ways that have been mentioned above. Popularizations of physics usually tell us that the ideas of Newton and Maxwell failed when they were applied on an astronomic or atomic scale, but remain perfectly good approximations to the phenomena of the middle-sized world. But in what sized world can matter be inert and not inert, and space empty and not empty? And the middle-sized world is characterized more than anything else by the presence of living things, which the atoms-and-void picture never had any hope of explaining, but only of explaining away. Shall we at least say, though, that we have learned that the world is not a cosmos? Let us listen to David Bohm:

> The theory of relativity was the first significant indication in physics of the need to question the mechanist order. . . . [I]t implied that no coherent concept of an independently existent particle is possible. . . . The quantum theory presents, however, a much more serious challenge to this mechanist order . . . so that the entire universe has to be thought of as an unbroken whole. In this whole, each element that we can abstract in thought shows basic properties (wave or particle,

> etc.) that depend on its overall environment, in a way that is much more reminiscent of how the organs constituting living beings are related, than it is of how parts of a machine interact. . . . [T]he basic concepts of relativity and quantum theory directly contradict each other. . . . [W]hat they have basically in common . . . is undivided wholeness. Though each comes to such wholeness in a different way, it is clear that it is this to which they are both fundamentally pointing. (*Wholeness and the Implicate Order,* London: Ark, 1983, pp. 173–76)

According to Bohm, it is only prejudice and habit that keep the evidence of the wholeness of things from being taken seriously. The contrary view is not just an opinion, but one of those fundamental ways of looking, thinking, and interpreting that permit us to have opinions at all and to decide what is and what is not a fact. To abandon the ground beneath our feet feels like violence, especially when no new authority is at hand to assure us that there is somewhere else for us to land. We tend to prefer to live with unreconciled dualities. Descartes notoriously makes the relation of mind and body a "problem." Newton speaks in the General Scholium to the *Principia* as though gravitation were incapable of explanation by physics, a supernatural element in the world. Leibniz speaks of two kingdoms, one of souls and one of bodies, as harmoniously superimposed (as in *Monadology* 79). Kant tells us that we are free, except insofar as our actions are part of the empirical world. We sometimes speak of biology as something unconnected with physics, as though what is at work in a tree or a cat is not nature in its most proper sense. We have had the habit so long that we consider it *natural* to regard ourselves, with our feeling, perception, and understanding, as an inexplicable eruption out of a nature that has nothing in common with us. Might a more coherent way be found to put together our experience? Perhaps it would be worthwhile to suspend, at least for a while, our notions of what can be and what it is for something to be, to try out some other way of looking.

A Non-Mathematical Physics

The world as envisioned in Aristotle's *Physics* is more diverse than the world described by mathematical physics, and we must accustom ourselves to a correspondingly richer vocabulary in order to read the

Physics. Motion means one thing to us, but irreducibly many kinds of thing when Aristotle speaks of it, and the same is true of *cause*. We tend to use *nature* as an umbrella-word, a collective name for the sum of things, while Aristotle means it to apply to whatever governs the distinct pattern of activity of each kind of being. Different English words could be used for these three ideas, to bring out what is distinctive in Aristotle's meaning, but here it seems best to keep the familiar words and push their limits beyond their prevalent current meanings. Nature, cause, and motion are the central topics of the *Physics,* and they first come to sight as questions; it is important to see that Aristotle and the later mathematical physicists were ultimately asking about the same things. Nature is mathematized not as an interesting game, or to abadon a harder task in favor of an easier one, but in order that the truth of it may be found.

In the *Assayer,* Galileo makes the famous claim that "this grand book, the universe, . . . is written in the language of mathematics." Later in the same book, in a discussion of heat, he explains why:

> I suspect that people in general have a concept of this which is very remote from the truth. For they believe that heat is a real phenomenon, or property, or quality, which actually resides in the material by which we feel ourselves warmed. . . . Without the senses as our guides, reason or imagination unaided would probably never arrive at qualities like these. Hence I think that tastes, odors, colors, and so on are no more than mere names so far as the object in which we place them is concerned, and that they reside only in the consciousness. Hence if the living creature were removed, all these qualities would be wiped away and annihilated. (*Discoveries and Opinions of Galileo,* New York: Anchor, 1957, pp. 237–38)

But shapes, sizes, positions, numbers, and such things are not mere names, imposed on objects by the consciousness of the living creature, because "from these conditions, I cannot separate [a material or corporeal] substance by any stretch of my imagination" (p. 274). The direct experience of the world has the taint of subjectivity, but the mathematical imagination captures the object just as it is. Sadder but wiser physicists today no longer try to read themselves out of physics; they know that they too are living creatures, interpreting the experience of a consciousness, with all the risk and uncertainty that accompanies such an activity. But our use of language may betray our second thoughts, and pull us back to Galileo's point of view.

What is motion? Do you think of something like a geometrical point changing position? What about a child moving into adolescence? Warmth moving into your limbs? Blossoms moving out of the buds on a tree? A ripening tomato, moving to a dark red? Are these other examples motions in only a metaphorical sense, while the first is correctly so called? Are the other examples really nothing but complex instances of the first, with small-scale changes of position adding up to large-scale illusions of qualitative change? For Aristotle, the differences among the kinds of motion determine the overall structure of the *Physics,* but they first of all belong together as one kind of experience. The kinds of becoming correspond to the ways being belongs to anything, and being-somewhere is only one aspect of being. A thing can also be of a certain size or of a certain sort or quality, and it can undergo motion in these respects by coming to be of another size or of a different quality by some gradual transition. It can even undergo a motion with respect to its thinghood. One thinks first of birth and death, but eating displays the same kind of motion. A cow chomps grass, and the grass is no longer part of the life of a plant, and is soon assimilated into the body of the cow. This is no mere change of quality, since no whole being persists through it to have first one, then some other quality belonging to it. Something persists, but first in one, then in another, kind of thinghood.

In any encounter with the natural world, it is the kinds of change other than change of place that are most prominent and most productive of wonder. Mathematical physics must erase them all and argue that they were never anything but deceptive appearances of something else, changing in some other way. Why? Because those merely local changes of merely inert bodies can be described mathematically. But if the testimony of the senses has a claim to "objectivity," and to be taken seriously, that is at least equal to that of the mathematical imagination, such a reduction is not necessary. And in fact the reduction of kinds of motion that is required is not just from four to one, but to less than one. Aristotle has considerable interest in change of place, but such a thing is possible only if there are places. Motion as mathematically conceived happens in space, and in space there are no places. Underneath the idea of motion that is prevalent today lurks this other idea, unexamined and taken on faith, that there is such a thing as space.

Aristotle twice argues that the idea of space, or empty extension, results only from the misuse of mathematics. His argument is the exact counterargument to Galileo's claim that ordinary people project their non-mathematical ideas onto the world. Aristotle says that the mathematician separates in thought the extension that belongs to extended bodies. (This is sometimes called "abstraction," but the word Aristotle uses is the ordinary word for subtraction.) There is nothing wrong with this falsification of things, which makes it easier to study what has been isolated artificially, so long as one does not forget that the original falsification took place. But some people do just that, reading this extension, which they have subtracted from bodies, back into the world as though the world were empty and somehow existed on its own prior to bodies. Once one has done so, one can, in the imagination, examine this "space" and determine all sorts of things about it. It is of infinite extent, for example, and since it is entirely empty, no part of it can have any characteristic by which it could differ from any other part. If our impulse, when thinking about motion, is automatically to give it a mathematical image, that is because we have presupposed that the ultimate structure of the world is space. But this supposition is laden with consequences and ought not to be adopted blindly. Aristotle says that one of the reasons physics cannot be mathematical is that the mathematician abolishes motion. Physics is the study of beings that move, and motion is a rich and complex topic, but within the constraints of "space" every form of motion disappears, except for one that is diminished out of recognition. If it is in space that our examination begins, nature will be nowhere to be found (but will survive as a mere name) because space is, from the beginning, a de-natured realm.

Conversely, without the imposition of the idea of space, nature can be understood as part of the true constitution of things, because motion in all its variety can be present. But since motion is not reduced to the pre-explained realm of mathematics, it is necessary to understand what it is. Aristotle says that, so long as we are ignorant of motion, we are ignorant of nature as well. But how can one give a rational account of motion? To assign it to some other genus would seem to make it a species of non-motion. In fact, two of Aristotle's predecessors, Parmenides and Zeno, had argued that motion is completely illusory. Parmenides argued that any attempt to say that there is motion must claim that what is-not also is. And Zeno, in four famous paradoxes preserved by Aristotle, tried to show that any description of motion involves self-contradiction of some kind. It would seem that motion has to be accepted as a brute fact of experience, from which explanations can begin, but which cannot itself be explained. But Aristotle, for the first and perhaps only time ever, did give motion a place not only in the world but in a rational account of the world, explaining it in terms of ideas that go deeper. The Parmenidean challenge is met by Aristotle primarily in the *Metaphysics,* where he shows that being must be meant in more than one way. His response to Zeno's challenge spreads over the whole of the *Physics,* and it is concentrated in his definition of motion.

Aristotle defines motion in terms of potency and being-at-work. In the first book of the *Physics* a preliminary analysis of change discovers that the ultimate explanatory notions available to the inquiry are form, material, and the deprivation of form. Material is described as that which, by its own nature, inherently yearns for and stretches out toward form. This should never be called matter, by which we mean something that stands on its own with a determinate set of properties (has weight, occupies space, preserves its state of motion in a straight line). What Aristotle means by material, on the contrary, is (1) not inert, (2) not necessarily tangible, (3) relative to its form, which may in turn be material for some other form, (4) not possessed of any definite properties, and (5) ultimately a purely "ideal" being, incapable of existing in separation, which would be rejected by any "materialist." Form, in turn, does not mean shape or arrangement, but some definite way of being-at-work. This is evident in Book II of the *Physics* and is arrived at by argument principally in the *Metaphysics,* VIII, 2. Every being consists of material and form, that is, of an inner striving spilling over into an outward activity. Potency and being-at-work are the ways of being of material and form.

The usual translations render *potency* as *potentiality,* which might suggest mere indeterminacy or logical possibility, which is never the sense in which Aristotle uses it. What is worse, though, is the rendering of *being-at-work,* and the stronger form of it used in the definition, *being-at-work-staying-itself,* as *actuality.* This has some reference, by way of Latin, to activity, but is a useless word that makes it completely impossible to get anything resembling Aristotle's meaning out of the definition. "The actuality of the potentiality as a potentiality" becomes a seventeenth-century joke, the ultimate example of the destruction of healthy common sense by pretentious gobbledygook. Does it refer to the actuality that belonged to the potential thing before it changed? That is not a motion, but something that precedes one. Does it refer to the actuality that exactly corresponds to the pre-existent potentiality? That is not a motion either, but something left when the motion ends. Does it mean, though it would have to be tortured to give this sense, the gradual transformation of a potentiality into an actuality? That at least could refer to a motion, but only by saying that a motion is a certain kind of motion. Perhaps it means that motion is the actuality of a potentiality to be in motion. This is surely the silliest version of them all, but respected scholars have defended it with straight faces. An intelligent misinterpretation of the definition was put forward by Thomas Aquinas, who took it to mean that the special condition of a thing in motion is to be partly actual while partly potential, and directed toward greater actuality of that same potentiality. But this account would not distinguish motion from a state of balanced equilibrium, such as that of a rock caught in a hand, still straining downward but prevented from falling any farther. Thomas's interpretation is subject to this ambiguity because it focuses on an instantaneous condition, a snapshot of a thing in motion, which is what an actuality is, but by no means what a being-at-work is.

What Aristotle said was that motion is the being-at-work-staying-itself of a potency, just as a potency. When an ongoing yearning and striving for form is not inner and latent, but present in the world just as itself, as a yearning and striving, there is motion. That is because, when motion is present, the potency of some material has the very same structure that form has, forming the being as something holding-on in just that particular motion. This does not mean that every motion is the unfolding of some being into its mature form; every such unfolding can fall short, overshoot, encounter some obstacle, or interact in some incidental way with some other being. It does mean that no motion of any kind would take place if it were not for those potencies that emerge of their own accord from beings. Motion depends on the organization of beings into kinds, with inner natures that are always straining to spill into activity. Only this dynamic structure of being, with material straining toward form and form staying at work upon material, makes room for motion that is not just an inexplicable departure from the way things are, but a necessary and intrinsic part of the way things are.

For example, consider the most uninteresting motion you can think of, say the falling of a pencil from the edge of a desk onto the floor. What is the potency that is at work, and to what being does it belong? The potency is *not* that of being at that spot on the floor, and the being that has it is not the pencil at all, since it is no genuine being. The potency at work here is that of earth to be down, or of the cosmos to sustain itself with earth at the center. No motive power belongs to the pencil as such, but it can move on its own because there is present in it a potency of earth, set free to be at work as itself when the obstacle of the desk is removed. And the motion is not defined by the position or state in which the pencil happens to end up, but by the activity that governs its course; the former is an actuality, but is not a being-at-work. Just as Newton's laws give a set of rules for analyzing any motion, Aristotle's definition directs us in a different way to bring the structure of any motion into focus: first, find the being, and then find the potency of it that the motion displays, or to which the particular motion is incidental. No motion, however random or incidental, gains entrance into the world except through the primary beings that constitute the world.

Aristotle sometimes argues about a body A that moves from B to C. Our first impulse may be to let A be represented by a point, the motion by a line, and B and C by positions. But Aristotle always has in mind an A with some nature, and a motion that may be from one condition to another rather than one place to another. Even if a motion from place to place is in question, those places would not be neutral and indifferent positions, but regions of the cosmos, which might or might not be appropriate surroundings for body A. The argument might be about something like continuity, so general that the particular B and C need not be specified, but it makes all the difference in the world that they represent motion in its fullest sense, as spelled out in the definition. The mathematized sort of motion, which can be fully depicted on a blackboard, is vulnerable to the kinds of attack present in Zeno's paradoxes. Motion as Aristotle understands it, constituted by potency and being-at-work, deriving its wholeness and continuity from a deeper source, overcomes those paradoxes. (The particular arguments will be looked at in the commentaries on the text.)

It is evident from this account of motion that material and form are understood as causes. The usual examples given for material and formal causes, an inert lump of bronze and a static blueprint, miss the point, namely, that material meets form half-way and that form is always at work. And material and form cause not only motions, but everything that endures. We tend to speak of causes as events that lead to other events, since that is the only kind of causality that remains possible in a mathematically-reduced world, but Aristotle understands everything that is the case as resulting from causes, and every origin of responsibility as a cause. Something called the "efficient" cause has been grafted onto Aristotle's account; it means the proximate cause of motion, like the bumping of billiard balls. Efficient cause is sometimes even used as a translation of one of Aristotle's four kinds of cause, but not correctly. Aristotle speaks of the external source of motion as one kind of cause, the *first* thing from which the motion proceeds. The incidental and intermediate links, which merely pass motions around without originating them, are not causes at all except in a derivative sense. All of Aristotle's causes stem from beings, and they are found not by looking backward in time, but upward in a chain of responsibility.

A fourth kind of cause, in most cases the most important one of all, is the final cause. It is often equated with purpose, but purpose is only one kind of final cause, and not the most general. A deliberate action of an intelligent being cannot be understood except in terms of its purpose, since only in achieving that purpose does the action become complete. The claim that final causes belong to nonhuman nature becomes ludicrous if it is thought that something must in some analogous way have purposes. What Aristotle in fact means is that every natural being is a whole, and every natural activity leads to or sustains that wholeness. His phrase for this kind of cause is "that for the sake of which" something does what it does or is what it is. Does rain fall for the sake of the crops that humans grow? No, but it does fall for the sake of the equilibrium of the cosmos, in which evaporation is counterbalanced by precipitation in a cycle of ever-renewed wholeness. That wholeness provides a stable condition for the flourishing of plants and of humans in lives and acts that come to completion in their own ways. Aristotle's "teleology" is just his claim that nothing in nature is a fragment or a chance accumulation of parts. To grasp the final cause of anything is to see how it fits into the ultimate structure of things.

But surely there are fragmentary things and chance combinations to be found around us. Aristotle finds it as strange that some thinkers deny chance altogether, as it is that others think chance governs everything. But from Aristotle's standpoint, even chance always points back to that which acts for the sake of something, since it results from the interference of two or more such things. Chance therefore represents not an absence of final causes, but an overabundance of them, a failure of final cause resulting from a conflict among final causes.

Because such incidental interactions lead to innumerable unpredictable chance results, nature is not a realm of necessity, but neither is it a realm of randomness, since the forms of natural beings govern all that happens. Aristotle speaks of the patterns of nature as present not always but "for the most part." His way of understanding the causes of things does not do violence either to the stability or to the variability of the world, but affirms the unfailing newness-within-sameness that we observe in the return of the seasons and the generations of living things. It offers an example of a physics that interprets causality without recourse to mechanical necessity or mathematical law. Both the collision of billiard balls and the co-variation of the two sides of an algebraic equation are too random in their beginnings and too rigid in their consequences to be adequate images of the natures we know.

The Shape of the Inquiry

It was mentioned above that all Aristotle's inquiries are dialectical. His writings have structures that are not rigid but organic, with parts that are whole in themselves but arranged so that they build up larger wholes. In the first book of most of his works he reviews what has been said by his predecessors. In Book I of the *Physics* that review is combined with a preliminary analysis of change, which concludes that change always implies the presence of some material that can possess or be deprived of form. This first analysis of everything changeable into form and material is then available as a starting point to approach any later question. Next comes the heart of the *Physics,* in Book II and the beginning of Book III, of which an account was given in the last section of this introduction. It begins with a definition of nature that has all the characteristics Aristotle attributes in I, 1 to proper beginnings: the discussion starts from what is familiar to us, it is clear in its reference but unclear in its meaning, and it takes its topic as a whole and in general, without separating out its parts or their particular instances. Since it defines nature as an inner cause of motion, the first task is to explore the meanings of cause and motion, not as words or logical classes, but through disciplined reflection on our experience. The result is a sharpened and deepened understanding of a way of encountering and interpreting the world. This is a more sustained use of the kind of analysis that took place in Book I, an analysis that dwells on a topic to unfold into clarity what was already present in an implicit and confused way.

A second kind of analytic reasoning begins after motion has been defined, a successive examination of conditions presupposed by the presence of motion in the world. This occupies the rest of Books III and IV. Zeno had taught everyone that motion presupposes infinity, and Aristotle turns first to this. He finds a non-contradictory way to understand the infinite divisibility of motion, but his conclusion that there is no infinite extended body is incomplete as it stands. It depends upon the claim that things have natural places, and so the topic of place must be explored next. Place is understood as a relation to the parts of the cosmos, but this topic in turn depends on the next topic, the void, since the exclusive array of places in the world results from the impossibility of void. The exploration of the idea of void completes this sequence, since the arguments against it stand independently. But motion also entails time, to which Aristotle turns next, finding that it is not in fact a presupposition but a consequence of motion. Time is found to result from a comparison of motions to one another, a comparison that can be carried out only by a perceiving soul. Like place, time is not a preexisting container and is not graspable by the mathematical imagination. Each of them is an intimate relation among beings, intelligible only when the particularity of this world is taken into account.

The last four books of the *Physics* take up the kind of cause and the kind of motion that are least central in Book II. The formal, material, and final causes of a living thing are internal to it and constitute its nature, and it has parents that are external sources of its motions of birth, development, and growth. But as Aristotle mentions at the end of II, 2, both other human beings and the sun beget a human being. All life is dependent upon conditions supplied by the cosmos, which seems to maintain itself primarily through cycles of local motion. Books V through VIII trace a complex argument up to the source of all change of place in the world. In its broadest outline, that argument is reminiscent of the structure of the *Metaphysics*. Though the *Metaphysics* is put together out of a large number of independent pieces, it has perhaps the clearest line of unifying structure of any of Aristotle's works. The meaning of being is pursued through four most general senses, to an eight-fold array of kinds of non-incidental predications, to its primary sense of thinghood, to the source of thinghood as form, to the meaning of form as being-at-work, to the source of all being-at-work in the divine intellect. It thus culminates in the discovery of the primary being that is the source of all being, and it gets there from the innocent question, Which of the meanings of being is primary?

A similar progressive narrowing of the meanings of motion takes place in the *Physics*. In Book III motion is said to be of four kinds: change of thinghood, alteration of quality, increase and decrease, and change of place. In Book V it is argued that motion properly understood is from one contrary to another, passing through intermediate states or conditions. But coming-into-being and destruction should be understood strictly as changes not to a contrary but to a contradictory condition, abrupt changes that have no intermediate conditions to pass through. Thus in a strict sense there are only three kinds of motion. But in Book VI it is argued that there is a certain discontinuity in every qualitative change. If something black turns white, it goes through a spectrum of intermediate shades, but it can be regarded as still being black until sometime in the course of the motion. In a change of quantity or place, once the thing is in motion it has departed from its initial condition, however much one might try to divide the beginning of the motion. So in the still stricter sense of being unqualifiedly continuous, there are only two kinds of motion. Finally, it is pointed out in Book VII that quantitative change must be caused by something that comes to be present where the changing thing is, so that it depends always upon a change of place prior to it, and it is argued in Book VIII that change of place is the primary kind of motion in every sense in which anything can be primary. The analysis goes one more step, to the primary motion within the primary kind, which is circular rotation. This is the most continuous of motions, so much so that it alone can be considered a simply unchanging motion.

Though the definition of motion in Book III applies to all motions, its application is most straightforward in the case of those motions most opposed to the primary kind, those that involve the greatest amount of change. Birth, development, and growth obviously unfold out of potencies that are present beforehand, and these changes point most directly to the inner natures of things that operate as formal, final, and material causes. But at the opposite extreme of the spectrum of change there is changeless circular motion. Because it moves without changing, it can be in contact with a completely unvarying cause. The last step of the inquiry in the *Physics* is the uncovering of a motionless first mover, acting on the cosmos at its outermost sphere. It is a source of local motion that not only holds the cosmos together, but contributes to the conditions of life by descending through the lower spheres, including that of the sun, to maintain the stable alternation of the seasons. Nature is thus seen as twofold, originating in sources of two kinds, the inner natures of living things and the cause holding together the cosmos as the outer condition of life. This is reflected in a bi-polar relation of motion and change, in which the ascending scale of motions (leading to the first external mover) is also the descending scale of changes (starting from the coming-into-being of new beings). The two-directionality of the scale is all-important. Aristotle does not reduce change to change of place, but rather traces it back, along one line of causes, to change of place. But the primacy of local motion in the cosmos does not abolish the primacy of the opposite kind of change, spilling over out of potency, which guarantees that even changes of place will be wholes, not vulnerable to the attacks of Zeno. The *Physics* has a double conclusion. In its final and deepest refutation of Zeno, it demonstrates the continuity rooted in potency as present in the limit of mere change of place, and this demonstration becomes one of the last steps in the argument that uncovers the motionless cause of motion.

Acknowledgments, and Notes on This Volume

The interpretation presented here has been stewing for almost thirty years, since my first college teacher, Robert Bart, opened my eyes to Aristotle's definition of motion in particular, and to the whole project of looking beneath and behind the presuppositions of modern science. Jacob Klein's "Introduction to Aristotle" is by far the best short introduction to Aristotle's writings; it was my first guide in the exploration of them. I strongly recommend that the reader find and read it in either Jacob Klein, *Lectures and Essays* (St. John's College Press, Annapolis, 1985), or the anthology *Ancients and Moderns,* edited by Joseph Cropsey (Basic Books, New York, 1964). Klein had heard Martin Heidegger lecture on Aristotle in the 1930s. This translation owes much to Heidegger's example of the possibility of reading Aristotle directly, not through the language of either the Latin tradition or the science of recent centuries. Heidegger suffers in translation almost as much as Aristotle does, but a good English version of his lectures on Book II, Chapter 1, of the *Physics* is cited earlier in this introduction. Heidegger is too ready to see form as presence-at-hand, uninvolved in the joining of things and emptying of one thing into another, and he is much too ready to talk about "the Greek idea of (whatever)," when discussing an insight that may have been achieved by only one or two thinkers, but as an antidote to the deadening effects of most commentary on Aristotle he is hard to beat.

This translation was a gleam in my eye for about fifteen years, until it was made possible by the generosity of St. John's College, the National Endowment for the Humanities, the Beneficial Corporation, and the Hodson Trust. Students and colleagues at St. John's have read drafts of it in classes and study groups. I am grateful for their conversation. Above all I am grateful for the encouragement given to me in this work, shortly before his death, by J. Winfree Smith. Whatever faults this translation may have, it had the merit of giving delight to that good man.

In the following translation the numbers from 184a to 267b refer to page and column in the standard two-column Bekker edition. The numbers between the Bekker page and column numbers allow the reader to match up the translation (approximately) with the lines of the Oxford Classical Text of Ross. I have followed Ross's main text or, occasionally, the variant readings provided in his notes. In the first paragraph of V, 3, for example, Ross has needlessly scrambled the text; my translation follows the manuscripts in everything but the placement of one sentence. The old Oxford translation by Hardie and Gaye makes the usual gibberish out of Aristotle's central vocabulary, but apart from that it is a thoughtful and graceful text. It was an invaluable aid to me in discovering the meaning of many words and phrases, while Ross's commentary was the source of a number of useful references. Ordinary parentheses in the text contain Aristotle's own parenthetical remarks; square brackets are used occasionally for my own insertions, when these go beyond repeating an antecedent of a pronoun. In one instance (at the end of IV, 8) curly brackets are used around a well-known passage that is not in the early manuscripts but appears in some late sources.

How to Use This Guidebook

The text is interspersed with running commentary. A superscript circle next to a word or phrase in the translation means that it is explained in the next section of commentary. The reader should begin by reading the brief note on the highlights of Aristotle's central vocabulary. For further study there is an extensive glossary, intended in part as a supplement to this introduction. Some parts of Aristotle's text that are rather technical and unnecessary to a first study of the *Physics* have been removed to the Appendix. Thus, the order of parts of this book forms a suggested order of study, but it is not the only order. Those who want to see where Aristotle's argument is going might read the sections of commentary before reading the parts of the text to which they refer. Those who want to absorb Aristotle's vocabulary in its wholeness might study the glossary first.

This translation is not intended to stand in place of Aristotle's inquiry in pursuit of nature, but to draw you closer to it. If what you find in the translation makes you want to go further, you should consider reading Aristotle's own Greek. His grammar is elementary, and his style is so repetitious that it does not take long to catch on to; the only difficulty in reading him is the concentration required to keep his pronouns straight. But if that route does not appeal to you, you can still join with Aristotle just by doing your own thinking about the questions he raises, in the light of the broadened and deepened array of possibilities he leads us to see.

Annapolis, Maryland
Summer, 1994

A Note on Aristotle's Central Vocabulary

The two ultimate ideas that govern Aristotle's thinking are *thinghood (ousia)* and *being-at-work (energeia).*

The primary fact about the world we experience is that it consists of *independent things (ousiai),* each of which is a *this (tode ti),* an enduring whole, and *separate (choriston),* or intact. Since thinghood is characterized by *wholeness (to telos),* the wholeness of each independent thing has the character of an *end (telos),* or *that for the sake of which (hou heneka)* it does all that it does. This doing is therefore the being-at-work that makes it what it is, since it is *what it keeps on being in order to be at all (to ti ēn einai).* Thus thinghood and being-at-work merge into the single idea of *being-at-work-staying-itself (entelecheia).*

It follows that *nature (phusis)* is not just a sum of bodies but is an activity, seen in the birth, growth, and self-maintenance of independent things and in the equilibrium of the parts of the cosmos. The cluster of central ideas in Aristotle's thinking is built on a few word roots that overlap in meaning: the *phu of phusis,* meaning birth and growth; the *erg* of *energeia,* meaning work; the *ech* of *entelecheia (enteles ēchein),* meaning holding-on in some condition (in this case completeness); and the *ēn* of *to ti ēn einai,* meaning *being* in the progressive aspect of that verb. This active, dynamic character is present in the very *material (hulē)* of each thing, as a *potency (dunamis)* spilling over into the activity that gives the thing its *form (eidos* or *morphē).*

A more extensive discussion of these terms and of their connections to the rest of Aristotle's vocabulary and to works other than the *Physics* may be found in the Glossary.

Book I

Beginnings

Chapter 1

184a Since, in all pursuits in which there are sources or causes or elements, it is by way of our acquaintance with these that knowing and understanding come to us (for we regard ourselves as knowing each thing whenever we are acquainted with its first causes and first beginnings, even down to its elements), it is clear that also for the knowledge of nature one must first try to mark out what pertains to its sources. On the other hand, the natural road is from what is more familiar and clearer to us to what is clearer and better known by nature; for it is not the same things that are well known to us and well known simply. For *20* this reason it is necessary to lead ourselves forward in this way: from what is *less* clear by nature but clearer *to us* to what is clearer and better known by nature. But the things that are first evident and clear to us are more-so the ones that are jumbled together, but later the elements and beginnings become known to those who separate them out from these. Thus it is necessary to proceed from what is general to what is particular, for it is the whole that is better known by perceiving, and what is general is a kind of whole since it embraces many *184b* things as though they were parts. Something of this same kind happens also with names in relation to their meanings, for a name too signifies some whole indistinctly, such as a circle, but the definition takes it apart into particulars. Children too at first address all men as father and women as mother, but later distinguish each of them.

Chapter 2

It is necessary that there be either one original being or more than one, and if one, either motionless, as Parmenides and Melissus say, or in motion, as the writers about nature say, some claiming that air is the first source, others that it is water. But if there are *20* more than one, they are either finite or infinite, and if finite but more than one, either two or three or four or some other number, or if infinite, either in such a way as Democritus says, one in kind but unlimited in shape, or differing in kind and even opposite. And they make a similar inquiry who inquire how many beings there are, for of those first things out of which beings are constituted, they inquire whether they are one or many, finite or infinite, so they are inquiring whether the original and elementary being is one or many.

185a Now to consider whether being is one and motionless is not to be examining nature. For just as it no longer belongs to the geometer to give an account to someone who rejects his starting points, but either to a different science or to one common to all knowledge, so is it with the one considering origins. For it is not any longer an origin if it is one only and there is therefore only one thing, for the origin is *of* something or some things. So to consider whether there is thus one thing is like discussing any other thesis whatever of people who argue for the sake of argument (such as the Heracleitean thesis,° or if someone should say that being is one man), or like resolving a debater's argument, which is exactly *10* what both arguments are, those, namely, of Melissus and Parmenides. For they take up things that are false, and their accounts are illogically put together. That of Melissus is more crude, and presents no impasse: once one is given one absurd thing° the rest follow, and this is in no way difficult. But for us, let it be assumed that the things that are by nature, either all or some of them, are in motion, which is obvious from examples. Nor, at the same time, is it appropriate to resolve all errors, but only as many as someone falsely concludes, demonstrating from first principles, and not those that are not of that kind; for instance, the squaring by means of segments belongs to the geometer to refute, but that of Antiphon° does not belong to the geometer. But even though they do not speak about nature, they *20* incidentally speak of things that are impasses in the study of nature. It is perhaps just as well to discuss them a little bit, for the examination is a philosophic one.

The starting point most appropriate of all, since being is meant in more than one way, is in what sense they mean it who say that everything *is* one. Is it that everything is an independent thing, or a so-much, or an of-this-kind, and in turn, what one thing is everything? Is it, say, one human being or one horse or one soul, or is this rather one of-this-kind, such as white or hot or something else of the sort? For all these things differ greatly, and each is impossible to claim. For if it were to be *both* an independent thing *and* an of-this-kind and so-much, and these either loosened from one another or not, beings would be many; but if everything is an of-this-kind or a so-much, then whether there is or *30* is not an independent thing, it is absurd, if one may call the impossible absurd. For none of the other senses of being is separate, other than thinghood, since everything is attributed to thinghood as what underlies it. But Melissus says that being is infinite. Therefore being is a so-much, since the infinite is in the genus of how much, while for an independent thing *185b* or a quality or a being-acted-upon to be infinite is not possible except incidentally, if at the same time some of them might also be so-much. For the articulation of the infinite makes use of the so-much, but not of thinghood nor of-this-kind. If, then, it is both an independent thing and a so-much, being is two and not one; but if it is an independent thing only, it is not infinite, nor will it have any magnitude, for then it would be a so-much.

Further, since also *one* itself is meant in more than one way, just as is being, one must examine in what way they mean that the whole is one. And what is said to be one is either the continuous or the indivisible or those things of which the articulation of what it is for *10* them to be is one and the same, such as mead and wine. Accordingly, if it is continuous, the one is many, for the continuous is infinitely divisible. (But there is an impasse about the part and the whole, though perhaps it does not connect with the argument but is just by itself: whether the part and the whole are one or more than one, and in what way one or more, and if more, in what way more, even about the parts of what is not continuous. And if each part is one with the whole as indivisible from it, the parts would also be one with one another.) But if it is one as indivisible, nothing would be either so-much or of-this-kind, and being would be neither infinite, as Melissus says, nor finite, as Parmenides does, since a limit *20* may be indivisible but not what is limited. But if all beings are one in meaning, like a robe and a cloak, they turn out to be asserting the Heracleitean account; for being-good and being-bad would be the same thing, and being-good and being-not-good—so that what is good and what is not good would be the same thing, as would a human being and a horse, and their account would not be about the being-one of all things but about the being-nothing of all things.

Even being of such-and-such a kind and being so-much in size would be the same.

And even the later ancient thinkers were disturbed that the same thing might become both one and many for them. For this reason some abolished "is," such as Lycophron, and *30* others restructured the language so that a human being "has whitened" rather than "is white" and "walks" rather than "is walking," in order not to make the one be many by attaching "is," as though *one* or *being* were meant in only one way. But beings are many either in meaning (as being-pale and being-educated are different, though the same thing is both, and thus the *186a* one is many) or by division, as with the whole and its parts. In this respect they were already at an impasse, and they granted that the one is many—as though it were not admissible for the same thing to be both one and many when these are not opposites, since there is what is one potentially or fully at work.

Chapters 3 and 4 are detailed discussions of the errors of Melissus, Parmenides, and Anaxagoras, using the vocabularies of those thinkers. Their content is unnecessary to Aristotle's own argument, which resumes in Chapter 5. Chapters 3 and 4 are omitted here, but are placed in the Appendix.

Chapter 5

Everyone makes contraries the original beings: those who say that the whole *20* is one and not moved (for even Parmenides makes hot and cold original beings, though he calls them fire and earth), and those who make them rare and dense, and Democritus the full and void, of which he says that the former is as being and the latter as non-being. He also says that things are by means of position, shape, and arrangement, but these are classes to which contraries belong: to position, up and down, before and behind, to shape, pointed and without corners, straight and round. That, then, everyone makes the original beings in some way contraries is clear, and reasonably so. For the original beings must not be dependent on one another nor on anything else, and all things must be dependent on them, *30* but this belongs to the first contraries: since they are first they are not dependent on anything else, and since they are contraries they are not dependent on one another.

But it is necessary to examine how this follows also from reasoning. But it must be understood first about all beings that none either acts by nature at random or is acted upon by any random thing, nor does anything at all come into being by chance from any chance thing, unless one takes what happens incidentally. For how could white come into being from something educated if the educated were not incidental to something non-white or *188b* black? Rather, white comes into being from what is not white, and not from every non-white thing but from the black or what is in-between, and educated from what is not educated, though not from every such thing but from the uneducated, or from something between the two if there is any. Nor is a thing transformed by destruction into the first chance thing, the white, say, into the educated, except when it happens incidentally, but the white is transformed by destruction into the non-white, and not into a random non-white thing but into something black or in-between, and in the same way the educated is transformed by destruction into the not-educated, and not into a random one but into the uneducated, or something between the two if there is any. And this is the case similarly with *10* other things, since also the things that are not simple but composite are in accord with the same account, but this result escapes our notice since there is no name for what is arranged in an opposite way. For necessarily everything concordant comes into being from what is discordant, and the discordant from what is concordant, and the concordant is transformed by destruction into the discordant, and this not any random one but one that is opposite. And it makes no difference whether one speaks of concord or order or composition, for obviously the same account holds. But surely also a house or a statue or anything else whatever comes into being in the same way: the house comes from what is not put together but from these things separated in this way, and the statue or any shaped *20* thing from the shapeless. And each of these things is in some respects a certain order and in others a certain composition. So if this is true, everything that comes into being would come from and everything that is destroyed be destroyed into either its opposite or what is between them. But the in-between things are derived from their opposites, as colors from white and black. So all things that come into being by nature either are opposites or are derived from opposites.

Up to this point, most of the others have been following along closely together, just as we said before. For all of them say that the elements and the things they call original *30* beings are contraries, and even though they lay it down without argument, they say it nonetheless, as though compelled by the truth itself. But they differ from one another in that some take hold of things that are more primary, others things that are more derivative, some of things that are better known by reason, others things that are better known by sense perception. (For some set down as causes of coming into being the hot and the cold, others the wet and the dry, and others the odd and the even or strife and friendship, and these differ from one another in the way that was said.) So they say things that are in a certain way the same as one another, but also different: different in just the way they seem to be *189a* to most people, but the same to the extent that they are analogous. For they take things that belong to the same series of corresponding things, since some of the contraries contain other ones. In this way what they say is the same and different, as well as better and worse, and some speak of what is better known by reason, others, as was said before, of what is better known by sense perception. (For what is general is known by reason, but what belongs to each thing is known by sense perception, since reason is of the general but sense perception of the particular.) For example, the great and the small belong to reason, but the rare and the dense belong to sense perception. *10* That, then, the starting points must be contraries, is clear.

Chapter 6

The next thing would be whether the starting points are two or three or more. They cannot be one, because contraries are not one, nor infinite, because what *is* would not be knowable, in addition to which there is one kind of oppositeness in any one genus, and thinghood is one particular genus, plus the fact that it is possible that things are derived from a finite number of sources, and it is better to trace them to a finite number, as Empedocles does, than to an infinite number. For he thinks he has accounted for everything every bit as much as

Anaxagoras does from infinitely many. Moreover, there are some contraries that are more primary than others, and others that come into being from one another, such as sweet and bitter or white and black, but it is necessary that the starting points always remain what they are.

20 That, then, they are neither one nor infinite, is clear from these things, and since they are limited in number, not to make them two only has a certain reasonableness. For one would be at an impasse about how density could be of such a nature as to do anything to rarity or it to density. And it is similar with any of the other oppositions whatever; for friendship does not bring strife together or make anything out of it, nor strife out of *it,* but both act on some other, third thing. And some assume even more things, out of which they construct the nature of things. But on top of these things one might produce this further impasse if someone were not to set down some nature different from the contraries, since *30* of no thing do we see the contraries constitute what it is, but the starting point should not be predicated of any underlying thing, for it would be a source of the source. For the underlying thing is a source, and seems to be prior to that which is predicated of it.

Further, we say that one independent thing is not contrary to another independent thing; how then could an independent thing be derived from what are not independent *189b* things, or how could something that is not an independent thing be prior to an independent thing? Therefore, if one were to regard as being true both the earlier argument and this one, it is necessary, if one is going to preserve them both, to set down some third thing, as those assert who say that the whole is some one nature such as water or fire or something between these. But the in-between seems better, for fire and earth and air and water are already entangled with oppositions. Not unreasonably then do those who make the underlying thing different from these do so, or do those who make it one of them make it air. For air least of them has sensible differences, and water next. But everybody decks out *10* this one thing with contraries, with density and rarity or with more and less. And plainly these are in general exceeding and falling short, as was said before. And this opinion seems to be an ancient one, that the one and exceeding and falling short are the sources of things, but not in the same way, but the older thinkers say that the two act and the one is acted upon, but some of the later ones say rather the reverse, that the one acts and the two are acted upon.

To say that the elements are three, based on these and other such things, would seem to the one who examines it to have some reason, as we said, but it is no longer reasonable to say they *20* are more than three. As for being acted upon, one thing is sufficient, and if among four things there were two oppositions, there would need to be present, separate from each pair, some other nature between them; and if the oppositions, though they are two, could have come into being from one another, one of them would be superfluous. At the same time, it is impossible that there be a greater number of first oppositions. For thinghood is one particular kind of being, so the starting points could only differ from one another in being more primary or more derivative, but not in kind. For always in one genus there is one opposition, and all oppositions seem to lead back into one. That, then, there is neither one element, nor more than two or three, is clear; but which of these is so, as we said, is a great impasse.

Chapter 7

30 On this point, let us, making an approach, speak first about all becoming, for it is a natural thing that, after speaking first about what is common, one examines in that way what is peculiar to each thing. For we say one thing comes into being from another or something from something different when we are speaking either of simple things or composite ones. I mean this in this way. For either a human being becomes educated, or *190a* the not-educated becomes educated, or the not-educated human being becomes an educated human being. By a simple thing that becomes something, I mean the human being or the not-educated, and by a simple thing that it becomes, the educated; but both what it becomes and the thing that becomes it are composite when we say the not-educated human being becomes an educated human being. Of one sort of these things, it is said not only that this thing comes into being, but also that it does so from that thing, as from the not-educated the educated, but this is not said in every case; for not from a human being did he become *10* educated, but a human being became educated. And of things that come into being in the way that we say simple things do, one sort becomes something while persisting, but another sort while not persisting; for the human being persists when he becomes educated and is still a human being, but the not-educated or the uneducated persist neither simply nor in composition.

Now that these things have been distinguished, from all coming into being there is this to be grasped, if one looks over it as we recommend: that something must always underlie the coming into being, and even though this is one in number, in form it is not one, and by "in form" I mean the same thing as in articulation. For it is not the same thing to be a human being and to be uneducated. And one thing persists but another does not persist; what is *20* not opposite persists (for the human being persists), but the not-educated or uneducated does not persist, nor what is composed of both, such as the uneducated human being. That something comes *from* something, rather than something becomes something, is said in the case of what does not persist, as the educated comes from the uneducated, but not from a human being; but even in the case of things that persist it is sometimes said that way. For we say a statue comes from bronze, not that the bronze becomes a statue. But what is from a non-persisting opposite is said in both ways, both from this comes that and this becomes that: for both from the uneducated comes the educated, and the uneducated becomes *30* educated. Hence it is also thus with the composite, for it is said both that from the uneducated human being comes the educated and that the uneducated human being becomes educated.

But becoming is meant in more than one way, and of some things it is meant not that it comes into being but that this thing becomes something else, but to come into being simply is meant only of independent things, while in the other cases it is clear that something must underlie what becomes. (For it belongs to some underlying thing that it becomes so-much or of-this-kind or in relation to something else or somewhere, because only thinghood *190b* is not predicated of something else which underlies it, but all the other [ways of being] are predicated of an independent thing.) But that independent things too, as well as whatever else simply is, come into being from some underlying thing, would become clear to those who examine them. For always there is something that underlies, out of which the thing comes into being, as do plants and animals from seed. And of the things that come into being simply, some come into being by change of shape, as does a statue, some by addition, such as growing things, some by subtraction, as does Hermes out of the stone, some by putting together, such as a house, and some by alteration, as do things that turn into something else on account of their material. All the things that come into being in these *10* ways obviously come from underlying things. So it is clear from what has been said that everything that comes into being is always composite: there is something that comes into being and there is something that becomes this, and this latter in two senses, either what underlies or what is opposite. I mean that the uneducated is opposite, and the human being underlies, and what is shapeless or formless or disordered is an opposite, but bronze or stone or gold an underlying thing.

It is clear then that if there are causes and sources of the things that are by nature, *20* from which first things they are and have come to be not incidentally but what each is said to be in virtue of its thinghood, then everything comes to be out of something underlying and form. For the educated human being is composed in some way out of human being and educated, since you could take it apart into the articulations of those things. It is clear then that what comes into being would come from these things. But while the underlying thing is one in number, it is two in kind (since the human being or the gold or in general the material is manifold, for it is primarily a *this,* and while it is not incidental that the thing that comes into being comes from it, still the deprivation or opposition is incidental to it). But the form is one, such as the order or the education or any of the other things that are predicated in this way. *30* Hence there is a way in which one must say that the starting points are two, and another in which they are three; and there is a way in which they are contraries, as if one were to speak of the educated and uneducated, or the hot and cold, or the concordant and discordant, but another in which they are not, since it is impossible for contraries to be acted upon by one another. But this also is resolved by the underlying thing's being something different, for this is not a contrary. So the starting points are in a certain way not more than the contraries, but two in number in this way of speaking, but neither are they altogether two on account *191a* of the being different from them of the underlying thing, but three. For being a human being is different from being uneducated, and being shapeless from being bronze.

How many are the starting points of the things involved in natural coming into being, and in what way they are so many, has been said, and it is clear that something must underlie the contraries and the contraries must be two. But in another way this is not necessary, for it would be sufficient if one of the contraries were to bring about the change by its absence and presence. And the underlying nature is knowable through analogy. For *10* as bronze is to a statue or wood to a bed or as the formless is before taking on its form, in relation to any of the other things that have form, so is this nature, in relation to an independent thing or a *this* or a being. This then is one starting point (though it *is* not one thing, nor *is* it at all in the same way as a *this),* and one starting point is the articulation that belongs to it, and further there is what is contrary to this, its deprivation. That these are somehow two and somehow more, was said above. It was said first that only the contraries were starting points, but later that something must also underlie them and that they must be three; but from what is being said now, it is clear what the difference between the contraries is, how the *20* starting points stand toward one another, and what the underlying thing is. But whether the thinghood of the thing is the form or what underlies it, is not yet clear. But that the starting points are three, and in what way three, and what their character is, is clear. So how many and what the starting points are have been brought into view by these means.

Chapter 8

After these things, let us explain that only in this way is the impasse to which the ancients came gotten past. For those who first inquired in a philosophic way into the truth and the nature of things got lost, as though they had been pushed aside into some other road by inexperience, and they say that none of the beings either comes into being or is destroyed, since it is necessary that what comes into being come either out of what is or *30* out of what is not, and out of both of these it cannot come; for a being would not come into being (since it already is), and from what is not, nothing could come into being, since *something* must underlie it. And building up the result successively in this way, they say that there are not even many things, but only being itself.

So they took hold of this opinion through what has been said. But we say that for something to come into being out of what is or what is not, or for what is not or what is to do something or be acted upon or become anything whatever to which one might point, is *191b* in one way no different than for a doctor to do something or be acted upon or be or become something out of being a doctor; so since this is meant in two ways, clearly so too is "from what is" or "what is acts or is acted upon." Now the doctor builds a house not as a doctor but as a housebuilder, and turns pale not as a doctor but as swarthy; but he heals or becomes a failure at healing as a doctor. But since we say most properly that a doctor does or suffers something or becomes something from a doctor when *as* a doctor he suffers or does or becomes these things, it is clear that also "this comes into being from what is not" *10* means what is not insofar *as* it is not. They left out this very thing, not having distinguished it, and on account of this mistake they made so great an additional mistake as to believe that nothing comes into being or is anything else, but rather to abolish all becoming. Now we and they say that nothing comes into being simply from what is not, but surely in some way a thing comes into being from what is not, for example incidentally. (For from its deprivation, which in virtue of itself is something that is not, and which does not continue to be present, something comes into being; but this is wondered at, and it seems impossible that something thus comes from what is not.) But likewise, neither does a thing come into being from what *is,* nor does what is come into being, except incidentally. But in this way *20* what is does come into being, in the same way as if animal were to come from animal or a certain animal from a certain animal, for example, if a dog were to come into being from a dog or a horse from a horse. For not only would the dog come into being from a certain animal, but also from animal, though not as animal, since this is still present. But if something is going to become an animal *not* incidentally, it will not be from an animal, and if something is going to become a being in this way, it will not be from a being. Nor will it be from what is not, for it was said that for us "from what is not" means insofar as it is not. Moreover, we do not abolish everything's either being or not being.

This then is one way to go, but another is that it is possible to mean these things in *30* respect to potency as well as to their being-at-work, but this distinction has been made in other places with more precision. So (as we said), the impasses are dissolved on account of which they were compelled to abolish some of the things mentioned; for on account of this they formerly detoured so much from the road to coming into being and passing away, and change in general. For this nature having been perceived, their entire mistake would have dissolved.

Chapter 9

There are some others° who have touched on it, but not sufficiently. For, *192a* first of all, they allow something to come into being simply out of what is not, on which point Parmenides speaks rightly; second, it seems to them that if something is one in number, it is also only one in potency. But this is very different. For we say that material and deprivation are different things, and of these the one is a non-being incidentally, namely the material, while the deprivation is so in its own right, and the one, the material, is almost, and in a certain respect is, an independent thing, which the other is not at all. But they make the non-being the great and the small alike, either both together or each separately. *10* So this triad and that one are of completely different characters. For they advanced as far as this, that there must be some underlying nature, but they make this one; for even if someone makes it a dyad, calling it great and small, nonetheless he makes them the same, for he overlooks the other nature. For the nature that persists is a co-cause with the form of the things that come into being, like a mother, while the other portion of the opposition might often be slandered as not being at all by one who fixes his thinking sternly upon it as upon a criminal. But since there is something divine and good and sovereign, we say that there is something opposite to it, and something else which inherently yearns for and *20* stretches out toward it by its own nature. For them, it follows that the contrary yearns for its own destruction. However, it is not possible either for the form to long for itself, since it is not defective, or for its contrary to long for it (since contraries are destructive of one another), but it is the material that does this, as does the female for the male or the ugly for the beautiful, except that in its own right it is neither ugly nor female, except incidentally.

There is a sense in which the material passes away and comes into being, and there is a sense in which it does not. As that in which a thing is, the material does in its own right suffer destruction (for that which is destroyed is in it, namely the deprivation), but as what is by way of potency, it does not in its own right suffer destruction, but is itself necessarily indestructible *30* and ungenerable. For if it were to come into being, there would have to be something underlying it, present all along, out of which it first came to be; but this is the nature itself of which we are speaking, so it would be before it came to be. (For by material I mean what first underlies each thing, out of which something comes into being, which is present all along, but not incidentally.) And if it were destroyed, it would arrive at this condition last, and so, before being destroyed [in its own right], it would have been destroyed [as a deprivation].

About the starting point in the sense of form, whether it is one or many and what it *192b* or they is or are, it is the work of first philosophy° to mark out with precision, so let it be set aside until that occasion. But we will speak of the forms of natural and destructible things in the elucidations that follow.

That, then, there are starting points, and what they are, and how many in number, let it have been marked out in this way for us, but starting over from another starting place, let us speak in a different way.

Commentary on Book I

The pre-Socratic philosophers speculated about nature, and they began at the beginning, with the ultimate beings that originate and govern all things. Aristotle begins the *Physics* with the reflection that, while the highest causes are the things that come first simply, a knowledge of them does not come first for us. Early modern philosophers, from the time of Descartes, as well as the most ancient Greek thinkers, attempted to secure all knowledge by reasoning only from the ultimate sources of things. Aristotle understands philosophy as inquiry in quest of those sources, which cannot help starting from where we find ourselves, with a general sense of how things are and an array of better and worse opinions.

The first book of each of Aristotle's inquiries reviews the opinions of previous thinkers. A prominent opinion of a respected thinker will contain either some valuable kernel of truth or some interesting error, and in either case it will point the inquiry in a direction. In Book I of the *Physics,* the most interesting error is the teaching of Parmenides that "all is one," since in denying manyness and, consequently, all motion and change, that teaching abolishes nature itself. Aristotle compares it to the claim of Antiphon, that, when rectilinear polygons of increasing numbers of sides are successively inscribed in a circle, one of them must eventually equal the area of the circle. Unlike the mistake of Hippocrates of Chios, who did find the areas of some curvilinear segments, and thought wrongly that he had found that of the circle, Antiphon's error is not within geometry at all but prior to it and blocking the way to it. Parmenides, seeking the permanent being behind the appearances of change, went so far that he left no way to reason back to nature. In its extremism, the Parmenidean thesis merges with its opposite, the claim of Heracleitus that "everything is in flux and nothing stays what it is," since both opinions collapse all distinctions. The refutation of Parmenides is primarily in the *Metaphysics*, where Aristotle shows that being is meant in more than one way, and even *one* must be understood in more than one sense. Here, at the end of Chapter 2, Aristotle points out that it is no contradiction to say that the same thing both is and is not, and is both one and many, in senses in which those attributes are not opposites, and thus he sufficiently secures the possibility of nature as a topic of inquiry.

Two of the most difficult and technical chapters in all Aristotle's works follow, with detailed refutations of various opinions of Parmenides, Melissus, and Anaxagoras. Melissus was a defender of Parmenides, but he was a shallow thinker. The "one absurd thing" (185a, 12) from which he reasons is the claim that whatever is everlasting must be infinite in size. Anaxagoras is chosen as asserting the opposite of the Parmenidean claim, by making the ultimate beings infinite in number. Aristotle's own argument resumes in Chapter 5, steering a middle course between these errors. It takes the form of a general analysis of change. Any change presupposes opposites, since no change is ever a random replacement of one condition by another; only the not-white can become white, and only the uneducated can become educated. But bare opposites such as white and not-white cannot turn into each other; some underlying thing that had one of the opposites in it comes to have the other in it. Now either of these opposites gives some sort of order or definiteness or *form* to the indeterminate underlying thing, and with respect to it, the other opposite is a lack or deprivation of that form. Deprivation of a form in some underlying thing capable of possessing that form is a special kind of non-being. Parmenides' thinking had lost its way over the impossibility of thinking what is not, while all change would have to be from what was not. But while a rock is not educated, only a human being can be uneducated. The deprivation of form in some underlying material to which that form is suited is a potent kind of non-being that yearns for what it lacks. The single ultimate opposition that stands behind all possible pairs of opposites is the distinction between form and material. Aristotle is often accused of taking this distinction from the realm of art and imposing it arbitrarily on nature. One needs only to read Chapters 5 through 9 of Book I of the *Physics* to see that this distinction is in fact deduced as a necessary condition of change in general.

The idea of form is explored in many of Plato's dialogues, and it is later members of Plato's Academy to whom Aristotle refers at the beginning of Chapter 9 as not having paid sufficient attention to the nature of material. Aristotle understands material as potency, stretching out toward form. An account of form itself belongs to "first philosophy," what we call metaphysics, and can be found especially in Books VIII and IX of Aristotle's *Metaphysics.* For the purposes of the study of nature, which Aristotle calls "second philosophy," form and material together, necessarily involved in all becoming, are available as starting points for an approach to understanding.

Book II, Chapters 1–3

Causes

Chapter 1

192b, 8 Of the things that are, some are by nature, others through other causes: by nature are animals and their parts, plants, and the simple bodies, such as earth, fire, air, and water (for these things and such things we say to be by nature), and all of them obviously differ from the things not put together by nature. For each of these has in itself a source of motion and rest, either in place, or by growth and shrinkage, or by alteration; but a bed or a cloak, or any other such kind of thing there is, in the respect in which it has happened upon each designation and to the extent that it is from art, has no innate impulse of change *20* at all. But in the respect in which they happen to be of stone or earth or a mixture of these, they do have such an impulse, and to that extent, since nature is a certain source and cause of being moved and of coming to rest in that to which it belongs primarily, in virtue of itself and not incidentally. (I say not incidentally because someone might himself become a certain cause of health in himself if he is a doctor. Still, it is not in the respect in which he is cured that he has the medical art, but it happens to the same person to be a doctor and be cured, on account of which they are also sometimes separated from each other.) And similarly with each of the other things produced: for none of them has in itself the source *30* of its making, but some in other things and external, such as a house and each of the other products of manual labor, others in themselves but not from themselves, as many as incidentally become causes for themselves.

Nature then is what has been said, and as many things have a nature as have such a source. And every thing that has a nature is an independent thing, since it is something that underlies [and persists through change], and nature is always in an underlying thing. According to nature are both these things and as many things as belong to these in virtue of themselves, as being carried up belongs to fire. For this is not a nature, nor does it have *193a* a nature, but it is something by nature and according to nature. What nature is, then, has been said, and what is something by nature and according to nature. *That* nature is, it would be ridiculous to try to show, for it is clear that among the things that are, such things are many. But to show things that are clear by means of things that are unclear is the act of one who cannot distinguish what is known through itself from what is known not through itself. (That it is possible to suffer this is not unclear, for someone blind from birth might reason about colors.) So it is necessary that the speech of such people be about names, while they have insight into nothing.

10 Now to some it seems that nature or the thinghood of things by nature is the first thing present in each which is unarranged as far as it itself is concerned; thus the nature of a bed would be wood, and of a statue, bronze. And Antiphon says that a sign of this is that, if someone were to bury a bed, and what rotted had the power to put up a sprout, it would not become a bed but wood, since what belongs to it by accident is the arrangement according to convention and art, while the thinghood of it is that which remains continuously even while it is undergoing these things. And if each of *these* things is in the same case in relation to something *20* else (as bronze or gold to water, and bones or wood to earth, and similarly with anything else at all), *that* would be the nature and thinghood of them. On account of which, some say fire, some earth, some air, some water, some say some of these, and some all of these to be the nature of things that are. For whatever from among these anyone supposes to be such, whether one of them or more, this one or this many he declares to be all thinghood, while everything else is an attribute or condition or disposition of these; and whatever is among these he declares to be eternal (since for them there could be no change out of themselves), while the other things come into being and pass away an unlimited number of times.

In one way then, nature is spoken of thus, as the first material underlying each of the *30* things that have in themselves a source of motion and change, but in another way as the form, or the look that is disclosed in speech. For just as art is said of what is according to art or artful, so also nature is said of what is according to nature and natural. We would not yet say anything to be according to art if it is only potentially a bed and does not yet have the look of a bed, nor that it is art, and similarly not in the case of things composed by nature. *193b* For what is potentially flesh or bone does not yet have its own nature, until it takes on the look that is disclosed in speech, that by means of which we define when we say what flesh or bone is, and not until then is it by nature. So in this other way, nature would be, of the things having in themselves a source of motion, the form or look, which is not separate other than in speech. (What comes from these, such as a human being, is not nature but by nature.)

And this form or look is nature more than the material is. For each thing is meant when it is fully at work, more than when it *is* potentially. Moreover, a human being comes about from a *10* human being, but not a bed from a bed. On this account, they say that not the shape but the wood is the nature, since if it were to sprout, it would become not a bed but wood. But if, therefore, this is nature, then also the form is nature, for from a human being comes a human being. And still further, the nature spoken of as coming into being is a road into nature. For it is not like the process of medicine, which is meant to be a road not into the medical art but into health, for it is necessary that the medical process be from the medical art and not into it. But not thus is nature related to nature, but the growing thing, insofar as it grows, does proceed from something into something. What then is it that grows? Not the from-which, but the to-which. Therefore nature is the form.

20 But form and nature are meant in two ways, for deprivation is a sort of form. But whether in the case of a simple coming-into-being there is or is not a deprivation and an opposite, must be looked into later.

Chapter 2

Now that nature has been marked off in a number of ways, after this one must see how the mathematician differs from one who studies nature (for natural bodies too have surfaces and solids and lengths and points, about which the mathematician inquires), and whether astronomy is different from or part of the study of nature. For if it belongs to the one who studies nature to know what the sun and moon are, but none of the properties that belong to them in themselves, this would be absurd, both in other ways and because those *30* concerned with nature obviously speak about the shape of the moon and sun and especially whether the earth and the cosmos are of spherical shape or not.

The mathematician does busy himself about the things mentioned, but not insofar as each is a limit of a natural body, nor does he examine their properties insofar as they belong to them because they pertain to natural bodies. On account of this also he separates them. For in his thinking they are separated from motion, and it makes no difference, nor do they become false by being separated. Those who speak of the forms also do this, but without being aware of it, for they separate the natural things, which are less separable than the *194a* mathematical ones. This would become clear if one should try to state the definitions of each of these things, both of themselves and of their properties. For the odd and even, and the straight and the curved, and further, number, line, and figure will be without motion, but no longer so with flesh, bone, or human being, but these are spoken of like a snub-nose and not like the curved. The more natural of the mathematical studies, such as optics, harmonics, and astronomy, also show this, for they in a certain way stand contrariwise to geometry. For *10* geometry inquires about a natural line, but not as natural, but optics about a mathematical line, not as mathematical but as natural.

But since nature is twofold, and is both form and material, we must consider it as though we were inquiring about what snubness is. As a result, such things will be neither without material, nor determined by their material. And in fact, since there are two natures, one might be at an impasse about which of them belongs to the study of nature. Or is it about that which comes from both? But if it is about that which comes from both, it is also about each of the two. Then does it belong to the same study or different ones to know each? If one looks to the ancients, it would seem to be about material (for only a little bit *20* did Empedocles and Democritus touch on form or the what-it-is-to-be of things). But if art imitates nature, and if it belongs to the same knowledge to know the form and the material to some extent (as it is the doctor's job to know health and also bile and phlegm, in which health is, and the housebuilder's to know both the form of a house and its material, that it is bricks and lumber, and in like manner with the rest), it would also be part of the study of nature to pay attention to both natures.

Further, that for the sake of which, or the end, as well as whatever is for the sake of these, belong to the same study. But nature is an end and a that-for-the-sake-of-which. (For of those things of which there is an end, if the motion is continuous, the end is both the last *30* stage and that for the sake of which; which induced the poet to say, absurdly, "He has his death, for the sake of which he was born." For not every last thing professes itself to be an end, but only what is best.) And the arts make even their material, some simply and others working it up, and we make use of everything there is as though it is for our sake (for we are also in some way an end, and "that for the sake of which" is double in meaning,° but *194b* this is discussed in the writings on philosophy). But the arts which govern and understand the material are two, one of using and one of directing the making. The art of using is for that very reason somehow directive of the making, but they differ in that the one is attentive to the form, the other, as productive, to the material. For the steersman recognizes and gives orders about what sort of form belongs to the rudder, but someone else about what sort of wood and processes it will come from. In things that come from art, then, we make the material for the sake of the work, but in natural things it is in being from the beginning. Further, material is among the relative things: for a different form, a different material.

10 To what extent is it necessary for the one who studies nature to know the form or whatness? Is it just as the doctor knows connective tissue or the metal-worker knows bronze, to the extent of knowing what each is for the sake of, even about those things which, while separate in form, are present in material? For both a human being and the sun beget a human being. But what manner of being the separate thing and the whatness have, it is the work of first philosophy to define.

Chapter 3

These things having been marked out, it is necessary to examine the causes, both what sort there are and how many in number. For since this work is for the sake of knowing, but we think we do not yet know each thing until we have taken hold of the why *20* of it (and to do this is to come upon the first cause), it is clear that we too must do this about both coming into being and passing away and about every natural change, so that, once we know them, we may try to lead back to them each of the things we inquire about.

One way cause is meant, then, is that out of which something comes into being, still being present in it, as bronze of a statue or silver of a bowl, or the kinds of these. In another way it is the form or pattern, and this is the gathering in speech of the what-it-is-for-it-to-be, or again the kinds of this (as of the octave, the two-to-one ratio, or generally number), and the parts that are in its articulation. In yet another it is that from which the first *30* beginning of change or of rest is, as the legislator is a cause, or the father of a child, or generally the maker of what is made, or whatever makes a changing thing change. And in still another way it is meant as the end. This is that for the sake of which, as health is of walking around. Why is he walking around? We say "in order to be healthy," and in so saying think we have completely given the cause. Causes also are as many things as come *195a* between the mover of something else and the end, as, of health, fasting or purging or drugs or instruments. For all these are for the sake of the end, but they differ from one another in that some are deeds and others tools.

The causes then are meant in just about this many ways, and it happens, since they are meant in more than one way, that the non-accidental causes of the same thing are also many, as of the statue both the art of sculpture and bronze, not as a consequence of anything else but just as a statue, though they are not causes in the same way, but the one as material and the other as that from which the motion was. And there are also in a certain way causes of one another, as hard work is a cause of good condition and this in turn is a cause of hard work, though again not in *10* the same way, but the one as end and the other as source of motion. Further, the same thing is a cause of opposite things. For the present thing is responsible for this result, and we sometimes blame it, when it is absent, for the opposite result, as the absence of the pilot for the ship's overturning, whose presence was the cause of its keeping safe. But all the causes now being spoken of fall into four most evident ways. For the letters of syllables and the material of processed things and fire (and such things) of bodies and parts of a whole and hypotheses of a conclusion are causes as that out of which, and while the one member of each *20* of these pairs is a cause as what underlies, such as parts, the other is so as the what-it-is-for-it-to-be, a whole or composite or form. But the semen and the doctor and the legislator, and generally the maker, are all causes as that from which the source of change or rest is, but other things are causes as the end or the good of the remaining ones. For that-for-the-sake-of-which means to be the best thing and the end of the other things, and let it make no difference to say the good itself or the apparent good.

The causes then are these and are so many in form, but the ways the causes work are many in number, though even these are fewer if they are brought under headings. For cause *30* is meant in many ways, and of those of the same form, as preceding and following one another. For example, the cause of health is the doctor and also the skilled knower, and of the octave the double and also number, and always comprehensive things in relation to particular ones. Further, there is what is incidental, and the kinds of these, as of the statue, in one way Polycleitus and in another the sculptor, because it is incidental to the sculptor to be Polycleitus. And there are the things comprehensive of the incidental cause, as if a *195b* human being were the cause of a statue, or generally, an animal. And also among incidental things, some are more remote and others nearer, as if the pale man or the one with a refined education were said to be the cause of a statue. And all of them, both those meant properly and those incidentally, are meant some as potential and others as at-work, as of building a house, either the builder or the builder building. And similarly to the things that have been said, an account will be given for those things of which the causes are causes, as of this statue or a statue or in general an image, and of this bronze or of bronze or in general of *10* material, and likewise with the incidental things. Further, things tangling these and those together will be said, such as not Polycleitus nor a sculptor but the sculptor Polycleitus.

Nevertheless, all these are six in multitude, but spoken of in a twofold way: there is the particular or the kind, the incidental or the kind of the incidental thing, and these entangled or spoken of simply, and all as either at-work or in potency. And they differ to this extent, that what is at-work and particular is and is not at the same time as that of which it is the cause, as this one healing with this one being cured or this one building with *20* this thing being built, but not always so with what is potential. The house and the house-builder are not finished off simultaneously.

And it is necessary always to seek out the ultimate cause of each thing, and in just the same way as with the others. (For example, a man builds because he is a builder, but is a builder as a result of the housebuilder's art; this, then, is the prior cause, and thus with everything.) Further, the kinds belong to the kinds and the particulars to the particulars (as a sculptor to a statue, but this sculptor to this statue.) Also the potencies belong to the potencies, but what is at work corresponds to what is being worked upon. How many then are the causes and in what way they are causes, let it have been marked out sufficiently for us.

Commentary on Book II, Chapters 1–3

Since Book I was preparatory for the inquiry into nature, Book II is its true beginning. Aristotle identifies nature as an "innate impulse of change" that not only sets things in motion but governs the course of those motions and brings them to rest. Only certain things have such inner sources of motion. A tentative list of natural beings is given in the first sentence of Book II, but it is corrected in the second paragraph. The *parts* of animals are not independent things, so while blood, say, or bone is natural, neither of them is that to which an inner source of motion belongs primarily, in virtue of itself. It is only the whole animal that has a nature, or a whole plant. Similarly, *fire* cannot properly be said to have a nature, since it is incapable of being a whole. Like blood and bone, fire, along with earth, air, and water, is only part of the whole being that has a nature. The ordered whole of the cosmos is the one independent thing in nature that is not an animal or plant.

What is the nature of a natural being? Almost all of Aristotle's predecessors would have located the nature of anything in the unorganized material underlying it. It is this material nature that would reassert itself if, in Antiphon's fantasy-experiment, a buried bed were to rot and put up a sprout. Aristotle begins by granting this materialist account of nature, but then modifies it in three steps: even if nature is material, it is *also* form; but nature is *more-so* form than material; and finally, nature *is* form. This is a characteristic way that Aristotle argues, and it shows why sentences from his writings cannot be casually plucked out of context. The materialist account has an initial plausibility, but in nature, when something develops or grows, it is the to-which of the process, not the from-which, that governs it. And Antiphon's example misses the point in any case; the bed is certainly not natural, but what underlies it is not undifferentiated wood, and this would be apparent by his own experiment. If anything sprouted it would not be unarranged wood, which human artisans produce by doing violence to trees, but an intact young olive tree, or cypress, or oak. When one adjusts one's perspective, both the from-which and the to-which of any process of growth are the same, and each is the form of a natural kind—oak tree in one case, human being in another.

But how should form be understood? Those of Aristotle's predecessors and contemporaries who were not materialists, the Pythagoreans and the members of Plato's Academy, understood the true natures of things to be mathematical. They too miss the mark, but on the opposite side, since they treat natural things as though they had no material. Natural things are not determined by their material, but they are composed of both material and form. The mathematician loses sight of motion altogether, in the rich sense of motion that natural beings display, which includes birth, growth, and qualitative change. Nor can the mathematical perspective recognize anything as inherently whole, and it therefore cannot see the wholeness for the sake of which natural beings do all that they do. Aristotle's understanding of nature is causal and therefore cannot be mathematical. It is a striking fact about the mathematical physics of our own times that its founders repeatedly and explicitly abandon the quest for causes. Galileo, Descartes, and Newton tell us that the investigation of causes in nature is useless, unnecessary, and inappropriate to the science. One legacy of the centuries during which mathematical physics has flourished is the virtual disappearance of the idea of cause as the origin of responsibility for the way something is.

According to Aristotle, the responsibility for anything can be traced back along four lines of investigation. First there is the material out of which something comes to be. This is a cause only according to the understanding of material achieved in Book I, Chapter 9, as something deprived of and stretching out toward form. Form is a cause in a second and more adequate way. It comes to sight in the *Physics* in three steps: in I, 5, as the source of orderliness in anything, and in II, 1, first as the "look disclosed in speech" *(to eidos to kata ton logon)*, and then as the source of the being-at-work of anything. The look disclosed in speech is that to which we pay attention when we recognize anything as belonging to a kind. This notion of form comes to Aristotle from his teacher Plato, but Aristotle develops the idea of form first into that which anything keeps on being in order to be at all, and thus into the activity that makes each thing what it is. Form *is* nature because it is the inner source of being-at-work in anything that has such a source. But a human being, for instance, is not just a source of its own motion, but can be, as a parent, the source of coming-into-being in another, or, as a legislator, the source of going-to-war by other people. This third kind of cause is not what came to be called the "efficient cause"; that refers to a proximate, incidental cause such as a push or a bump, in which Aristotle has no interest. Causes in Aristotle's sense always stand at the beginning of chains of responsibility, so that they can be adequate answers to the question, Why? The most complete answer to that question is always found in the fourth kind of cause, the end or completion for the sake of which anything does anything. This has two senses, described by Aristotle in other places as the "for which" and the "of which." Sometimes animals act for the sake of their offspring, or wheat seeds are ground up to make bread for humans, but all such instances of external ends presuppose internal ends prior to them. The children for whom parents sacrifice must have lives in which they find completion, the kernels of wheat must be nutritive for the wheat plant itself if they are to be of any use to other species, and so on. Aristotle's "teleology" does not impose the human idea of purpose onto non-human nature, but recognizes that all natural beings are whole and act so as to preserve that wholeness and fulfill its potencies. Final causality governs the action of formal causes, and thus characterizes the whole realm of nature. This topic leads to the next part of the inquiry.

Book II, Chapters 4–9

Chance and Necessity

Chapter 4

195b, 30 Fortune and chance are spoken of among the causes, and many things are said to be and to come about through fortune or through chance. In what way, then, fortune and chance are among these causes, and whether fortune and chance are the same or different, *196a* and in general what fortune and chance are, must be examined. For some people are at an impasse even about whether they exist or not. For they say nothing comes about from fortune, but that, for everything whatever that we say comes from chance or fortune, there is some definite cause; for example, of coming by fortune into the marketplace and catching up with someone whom one wanted, but did not expect, to find, the cause is wanting to go to use the marketplace. And similarly with the rest of the things said to be from fortune, there is always something to take as the cause, but not fortune, since if fortune were anything it would seem strange, and truly so; and one might find it impossible to understand why in the world none of the ancient wise men, speaking about the causes of coming-to-be *10* and passing away, demarcated anything about fortune, but as it seems, they too regarded nothing as being from fortune. But this too is to be wondered at. For many things both come about and are from fortune and from chance, which everyone, though not ignorant that it is possible to refer each of them back to some cause of its coming about, which the earlier argument declared to be the abolition of fortune, nevertheless says are from fortune, some of them, though some are not. For this reason they were obliged to make some mention of it in some way. But surely they did not regard fortune to be any of those things such as friendship and strife, or intellect, or fire, or anything else of that sort. It is strange, then, *20* either if they did not acknowledge it to be, or if, supposing it, they left it aside, despite even sometimes making use of it, as Empedocles says that the air is not always separated in the highest place, but however it falls out. Certainly he says in his cosmogony that "it fell out thus as it was flowing at one time, but often otherwise." And most of them say that the parts of animals have come into being from fortune.

There are some who make chance responsible for this cosmos and all worlds. For they say that by chance there came about a vortex and a motion of separating out and settling into this arrangement of the whole. And this itself is in fact mightily worth *30* wondering at. For they are saying that animals and plants neither are nor come to be by fortune, but that either nature or intellect or some other such thing is the cause (for what comes into being from each seed is not whatever falls out, but from this one an olive tree, from that one a human being), but that the heavens and the most divine of visible things have come from chance, and there is in no way such a cause as there is of the animals and *196b* plants. But if this is the way things are, this itself is worth bringing one to a stop, and it would have been good for something to have been said about it. In this respect as well as others, what is said is strange, and it is stranger still to say these things when one sees nothing in the heavens happening by chance, but many things falling out by fortune among the things not assigned to fortune, though it would surely seem that the opposite would happen.

There are others to whom it seems that fortune is a cause, but one not disclosed to human understanding, as though it were something divine and more appropriate to miraculous agency. So it is necessary to examine what each is, and whether chance and fortune are the same or different, and how they fall in with the causes that have been marked out.

Chapter 5

10 First, then, since we see some things always happening in a certain way, and others for the most part, it is clear that of neither of these is fortune or what comes from fortune said to be the cause, neither of what is out of necessity and always, nor of what is for the most part. But since there are other things besides these that happen, and everyone says that they are from fortune, it is clear that fortune or chance is in some way. For we know that things of this kind are from fortune and that things from fortune are of this kind.

Now of things that happen, some happen for the sake of something and some not (and of the former, some in accordance with choice, some not in accordance with choice, but *20* both are among things for the sake of something), so that it is clear that even among things apart from what is necessary or for the most part, there are some to which it is possible that being for the sake of something belongs. And for the sake of something are as many things as are brought about from thinking or from nature. But whenever such things come about incidentally, we say that they are from fortune. (For just as a thing is something either in virtue of itself or incidentally, so also is it possible to be a cause, as of a house, the cause in virtue of itself is the builder's art, but an incidental one is the pale or educated man; the cause in virtue of itself, then, is definite, but the incidental one indefinite, for to one thing, infinitely many things incidentally belong.) Just as was said, then, whenever this happens among things happening for the sake of something, in that case it is said to be from chance or from fortune. (The difference between these in relation to one another is something that must *30* be distinguished later; for now let this be clearly seen, that both are among the things for the sake of something.) For example, someone gathering contributions would have come for the sake of collecting money, if he had known; but he came not for the sake of this, but it happened to him incidentally to go and to do this. And this was not through frequenting *197a* the place for the most part or out of necessity, but the end, the collection, though not belonging to the causes in him, is among choices and things that result from thinking. And in this case he is said to have come by fortune, but if he had chosen to, and for the sake of this, or if he frequented the place always or for the most part, not by fortune. It is clear then that fortune is an incidental cause among things proceeding from choice, which in turn are among those for the sake of something. Whence thinking and fortune concern the same thing, for there is no choice without thinking.

It is necessary, then, that the causes be indefinite from which what arises from fortune *10* comes about. Whence fortune too seems to be indefinite, and obscure to humans, and it is possible for it to seem that nothing comes about from fortune. For all these things are said correctly, reasonably. That is, there are things that come about from fortune: they come about incidentally, and fortune is an incidental cause, but of nothing is it the cause simply. As of a house, the cause is the builder's art, but incidentally a flute-player, also of coming to collect money when one has not come for the sake of this, the multitude of causes is unlimited. One wanted either to see someone, or look for someone, or get away from someone, or see a show. It is even correct to say that fortune is something non-rational, for *20* a reasoned account belongs to what happens either always or for the most part, and fortune is among things that come about outside of these ways. Thus, since causes of this kind are indefinite, fortune too is indefinite. Still, in some situations, one might be at a loss whether the things that happen to occur would become causes of fortune, as of health, the wind or the sun's warmth, but not having had a haircut. For among incidental causes, some are nearer than others.

Whenever something good turns out, fortune is called good, or indifferent when it is something indifferent, but good fortune or ill fortune are only spoken of when these outcomes are of some magnitude. For this reason too, to come within a hairsbreadth of obtaining some great evil or good is to be fortunate or unfortunate, because our thinking *30* picks out what comes out, but seems to hold off what was within a hairsbreadth as nothing. Further, good fortune is unstable, and reasonably, for fortune is unstable, since it is not possible for any of the things from fortune to be always or for the most part. Both, then, are causes, incidental ones as was said, both fortune and chance, among those things which admit of coming into being neither simply nor for the most part, belonging in turn to those which could come about for the sake of something.

Chapter 6

They differ because chance is more extensive, for everything from fortune is *197b* from chance, but not everything from it is from fortune. For fortune and what comes from fortune are present to beings to whom being fortunate, or generally, action, might belong. For this reason also, fortune is necessarily concerned with actions. (A sign is that good fortune seems to be either the same thing as happiness or nearly so, while happiness is a kind of action, namely doing well.) So whatever cannot act cannot do anything as a result of fortune either. And for this reason no inanimate thing nor any animal or small child can do anything as a result of fortune, because they do not have the power to choose in advance. *10* Neither good fortune nor misfortune belongs to them, except metaphorically, as Protarchus says the stones out of which altars are made are fortunate, because they are honored, while their quarry-mates are trampled on. But it belongs even to these things to be affected by fortune, whenever an active being acts on them in some way that results from fortune, but in no other way. Chance, though, belongs to the other animals and to many inanimate things, as we say a horse came along by chance, because he was saved by his coming, though he did not come for the sake of the being saved. Or a tripod fell down by chance, for it stood there in order to be sat upon, but did not fall down in order to be sat upon. So it is clear, among things happening for the sake of something simply, whenever they happen not *20* for the sake of what turned out, of which the cause was external, we then say that they are from chance. And as many of these things that happen by chance as are choices, and happen to those having the power of choice, are from fortune.

A sign of this is the phrase "in vain," which is said whenever what is for the sake of another thing does not come to pass for the sake of that, as, if taking a walk were for the sake of evacuation, but when one had walked this did not happen, we would say that having walked was vain and taking a walk futile, as though this was the "in vain": for something by nature for the sake of another thing, the not being brought to accomplishment of that for the sake of which it was and to which it was naturally disposed, since if someone said he had bathed in vain because the sun was not eclipsed, it would be ridiculous, for this was not for *30* the sake of that. Thus, even from its name, chance *[to automaton]* is that which itself happens in vain *[to auto matēn]*. For the stone fell not for the sake of knocking someone out; therefore the stone fell by chance, because it could have fallen by some agency and for the sake of the knocking out.

Chance is separated most of all from what comes from fortune in things that happen by nature. For whenever something happens contrary to nature, we say that it happened not by fortune but by chance. But it is also possible that this is for a different reason, since of the one the cause is outside, of the other inside.

198a What, then, chance and fortune are has been said, and how they differ from each other. And of the ways of being a cause, both of them are among things from which the source of motion is; for they are always among either things in some way by nature, or causes that come from thinking, but the multitude of them is infinite. But since chance and fortune are causes of things for which either intelligence or nature might have been responsible, whenever something incidentally becomes responsible for these same things, but nothing incidental is prior to things in virtue of themselves, it is clear that neither is the *10* incidental cause prior to what is in virtue of itself. Therefore chance and fortune are subordinate to intelligence and nature, so that if chance is responsible for the heavens as much as possible, it is necessary that intelligence and nature have a prior responsibility, not only for many other things, but also for this whole.

Chapter 7

That there are causes, and that they are as many in number as we said, is clear, for the why includes so many in number. For the why ultimately leads back either to the what-it-is, among motionless things (as in mathematics, for it ultimately leads back to the definition of straight or commensurable or something else), or to the first source of motion (as, Why *20* did they go to war? Because they were plundered), or something for the sake of which (in order to rule), or, among things that come into being, the material.

That the causes, then, are these and this many, is clear; and since there are four causes, it belongs to the one who studies nature to know about all of them, and he will supply what is due in the way of natural inquiry by tracing back the why to them all: the material, the form, the mover, and that for the sake of which. But often three of them turn back into one, for the what-it-is and that for the sake of which are one, and that whence the motion first is, is the same in form with these; for a human being brings forth a human being, and in general, as many things as, being moved themselves, cause motion, are the same in form with the things moved. (Whatever is not like this does not belong to the study of nature. For it causes motion not by having motion or a source of motion in itself, but being motionless. On which account there are *30* three studies, one about motionless things, one about things moved but indestructible, and one about destructible things.) So they supply what is due by tracing back the why to the material, and to the what-it-is, and to the first mover. About coming into being, they examine the cause mostly in this way: what comes about after what, and what did it do first, or how was it acted upon, and so on always in succession.

The sources which bring about motion naturally are twofold, of which one kind is *198b* not natural, for sources of that kind do not have in themselves a source of motion. And of this kind is whatever causes motion without being moved, as does not only what is completely motionless and the first of all beings, but also the what-it-is or form, for it is an end and that for the sake of which. So, since nature is for the sake of something, it is also necessary to know this, and one must supply the why completely: for example, that from this necessarily comes that (from this either simply or for the most part), *and* that if it is going to be, this will be (as from the premises, the conclusion), *and* that this is what it is for it to be, and because it is better thus, not simply, but in relation to the thinghood of each thing.

Chapter 8

10 One must say, first, why nature is among the causes for the sake of something, then, about the necessary, how it holds a place among natural things. For everybody traces things back to this cause, inasmuch as, since the hot and the cold and each thing of this kind are by nature a certain way, these things are and come into being out of necessity. For even if they also speak of another cause, they send it on its way after only so much as touching on it, one on friendship and strife, another on intellect.

Here is an impasse: what prevents nature from doing things not for the sake of anything, nor because they are best, but just as Zeus rains, not in order that the grain might grow, but out of necessity? (For it is necessary that what is taken up be cooled, and that *20* what is cooled, becoming water, come down; when this happens, growing incidentally happens to the grain.) Likewise, if the grain is ruined on the threshing-floor, not for the sake of this did it rain, to spoil it, but this was incidental. So what prevents the parts in something that is by nature from being the same way, say the teeth growing with the front ones sharp out of necessity, suitable for tearing, but the molars flat and useful for grinding the food, although not happening for the sake of this, but just falling together? Likewise with the other parts, to however many being for the sake of something seems to belong, *30* wherever everything happened to come together just as if it had been for the sake of something, these were preserved, having been put together advantageously by chance. Anything that is not like that has perished and still perishes, just as Empedocles says of man-headed offspring of cattle.

The account, then, by means of which one might come to an impasse, is this one or any other that might be of this kind; but it is impossible for things to be this way. For these things and all things that are by nature come about as they do either always or for the most part, but none of the things from fortune or chance do. For it does not seem to be from *199a* fortune or by coincidence that it rains often in winter, but it does if this happens in the dog days, nor scorching heat in the dog days, but in winter. If, then, it seems that something is either by coincidence or for the sake of something, and if things by nature cannot be by either coincidence or chance, they would be for the sake of something. But surely such things are all by nature, as even those making these arguments would say. Therefore, there is being-for-the-sake-of-something among things that happen and are by nature.

Further, among all things that are for some end, it is for the sake of this that what precedes it in succession is done. Accordingly, in the way that one performs an action, so *10* also are things by nature, and as things are by nature, so does one perform each action unless something interferes. But one acts for the sake of something, and therefore what is by nature is for the sake of something. For example, if a house were something that came into being by nature, it would come about in just the way that it now does by art, and if the things by nature were to come about not only by nature but also by art, they too would come about in exactly the same way as they do by nature. Therefore each is for the sake of another. And in general, art in some cases completes what nature is unable to finish off, but in others imitates nature. If then, what comes from art is for the sake of something, it is clear that what comes from nature is too, for the series of things from art *20* and from nature are alike, each to each, in the way that the later things are related to the earlier.

This is clear most of all in the other animals, which do nothing by art, inquiry, or deliberation; for which reason some people are completely at a loss whether it is by intelligence or in some other way that spiders, ants, and such things work. But if we move forward little by little in this way, it becomes apparent that even in plants what is brought together comes about in relation to the end, as the leaves for the sake of protection for the fruit. So if both by nature and for the sake of something the swallow makes a nest and the spider a web, and the plants make their leaves for the sake of their fruit, and their roots not *30* upward but downward for the sake of nourishment, it is clear that there is such a cause in things that come into being and are by nature. And since nature is twofold, both material and form, and the latter is an end but the former is for the sake of an end, the form would be the cause for the sake of which.

Now missing the mark happens even among things done according to art (for the grammarian on occasion writes, or the doctor gives out a drug, incorrectly), so it is clear that *199b* this is possible also among things done by nature. But if there are some things according to art in which what is done correctly is for the sake of something, but in the ones that miss the mark what is done is for the sake of something that is attempted but missed, it is the same among natural things, and monsters are failures of that for the sake of which they are. Therefore even the cattle-offspring, in their original constitution, if they were not able to come to some limit and end, would have come into being when some originating cause in them was disabled, as might happen now with the seed. Still it is necessary that a seed come into being first, and not straightaway the animals; "first the mixed-natured" was the seed. *10* That for the sake of which is also present in plants, though less articulated. So then did there come about among the plants, like the man-headed offspring of cattle, also olive-headed offspring of grapevines, or not? It would be strange, but it would be necessary, if it also happened among animals.

In general, it would be necessary, among seeds, that whatever chanced come into being, but the one speaking this way abolishes nature and what is by nature. For by nature are as many things as, moved continuously by some source in themselves, reach some end; from each beginning does not come the same end for them all, nor just what chances, but each always reaches the same end unless something interferes. That for the sake of which, and that which is for the sake of this, might also happen by fortune, as we say that a stranger *20* came by fortune and, having paid the ransom, went away, when he acted as though having come for the sake of this, but did not come for the sake of this. And this is incidental (for fortune is among the incidental causes, just as we said before), but whenever this happens always or for the most part, it is neither incidental nor by fortune. But among natural things, things happen always in the same way, unless something interferes.

It is absurd to think that a thing does not happen for the sake of something if we do not see what sets it in motion deliberating. Surely even art does not deliberate. If shipbuilding were present in wood, it would act in the same way as nature does, so if being *30* for the sake of something is present in art, it is also present in nature. This is most clear when someone practices medicine himself on himself; for nature is like that. That, then, nature is a cause, and in this way, for the sake of something, is clear.

Chapter 9

Does what is by necessity belong to things conditionally or simply? For now *200a* people suppose that what is by necessity is in the coming into being of things, as if someone were to think that the wall of a house came into being by necessity, because the heavy things are of a nature to be carried downward and the light ones on top, so that the stones and foundations are at the bottom, the earth above on account of its lightness, and at the very top the wood, since it is lightest. But even though it did not come into being without these things, it surely did not do so as a result of them, except as by means of material, but rather for the sake of enclosing and sheltering certain things. And similarly with everything else, in whatever being-for-the-sake-of something is present, each thing is neither without things *10* having necessity in their nature, nor as a result of them other than as material, but for the sake of something. For example, why is a saw thus? In order to do this and for the sake of this. But this which it is for the sake of would be incapable of coming about if it were not made of iron. It is necessary, therefore, that it be of iron if the saw and its work are to be. So the necessary is conditional, unlike the end. For the necessary is in the material, but that for the sake of which is in what is grasped in speech.

The necessary is present both in mathematics and in things that come about by nature, in ways closely resembling one another. For since the straight is a certain way, it is necessary that the triangle be equal to two right angles, but not the former because of the latter, despite the fact that were this not so, neither would the straight be as it is. But in *20* things that come to be for the sake of something, contrariwise, if the end is to be or is, then what precedes it will be or is; and if not, just as there [in mathematics] the first principle will not be so when the conclusion is not, so also here [in nature] with the end and that for the sake of which. For this too is a starting point, not of action but of reasoning (and of reasoning there; for there are no actions). So that if a house is to be, it is necessary that there come into being or be present or in general be these things as material for the sake of something, such as bricks and stones if it is a house. Nevertheless, the end is not present as a result of these, other than as material, nor will it be, just because of them. In general, however, neither the house nor the *30* saw will be if the bricks, in the former case, and the iron in the latter, are not; for neither will the starting points be the case there if the triangle is not two right angles.

It is clear that the necessary in natural things is the so-called material and its motions. And both must be stated as causes by the one who studies nature, but more so that for the sake of which. For this is responsible for the material, but the material is not responsible for the end. And the end is that for the sake of which, and the beginning comes from the definition and that which is grasped in speech; just as in things that come from art, since the *200b* house is such, these things must come into being or be present necessarily, and since health is such, *these* things must come into being or be present necessarily—so also if a human being is such, these things, but if these, these others in turn. Perhaps the necessary is even in the definition. For the work of sawing having been defined as a certain kind of dividing, this will not be unless it has teeth of a certain kind, and these will not be of that kind unless they are of iron. For even in the definition there are certain parts, as material of the definition.

Commentary on Book II, Chapters 4–9

Aristotle's four kinds of cause apply to everything that is by nature, and analogies to them apply to products of human art. But are there not some things that happen with no end or form governing them, just by chance or by mere necessity? Aristotle repeatedly remarks on the strangeness of everything said about chance by his predecessors, who either give it the ultimate responsibility for the world or abolish it altogether, or even do both. This combination is present in much of the thought of our times as well, in which the world itself (compare 196a, 27, with Chapter 6 of Descartes' *Le Monde)* and all living things (compare 198b, 29, and the other reference to Empedocles at 199b, 9, with Darwin's *Origin of Species*) are said to come about by chance but then run along in constant ways determined rigidly by necessity. Even the idea of "laws of nature," expressed in algebraic equations, implies that everything is governed by chance and necessity together, since values are plugged in to one side of the equation arbitrarily, and ground out on the other side necessarily.

Aristotle sees chance and necessity as missing the realm of nature altogether, on opposite sides. Natural beings and events display constant patterns that cannot be the result of chance alone, but they never display the machine-like uniformity of necessary determinism. Nature looks to the whole being or the whole activity, with a flexibility about parts, means, and other incidental attributes. In nature things happen in certain ways "for the most part." The world contains many things of many kinds, not running their course in isolation but interacting in ways that are literally innumerable. When two or more lines of causes cross, the result is incidental or accidental. Incidentally the housebuilder is a flute-player; accidentally the rock fell on someone's head. But each thing must first be something and act in accordance with what it is, before it can interact by chance with anything else. Chance is a genuine and widespread part of the world, but a subordinate one, and the mysterious thing called fortune is just chance that befalls human beings.

The famous example of the rain that spoils the crop that has already been harvested is widely discussed but rarely understood. This is Aristotle's own fault, since he leaves so much of the argument inexplicit. As with Antiphon's bed in II, 1, where we had to see for ourselves that whole trees of definite kinds, and not mere wood, are the only things that could sprout from rotting wood, in the way that human beings come from human beings, the crucial point about rain is left for us to supply (or is so obvious to Aristotle's students that it is taken for granted). The phrase "Zeus rains" is an idiomatic expression taken from Homer, but if we believe that rain happens by nature then we must recall the definition of nature. What is the being to which the activity of rain belongs *primarily*, as the result of an internal source of motion? The whole being is not clouds or air or water, each of which remains by nature in a certain region of the cosmos, but it is the cosmos as a whole; in the whole, evaporation and rain always balance out in the long run, and are governed by the yearly course of the sun up and down the sky. Like everything else in nature, rain is activity for the sake of an end, but the end is the maintenance of the cosmos. Rain is an activity of the cosmos. Rain also contributes to the growth of crops, but the relation between rain and any particular wheat-field is subject to a great deal of chance, since fostering wheat is incidental to the end that belongs to the cosmos. Rain is for the sake of crops in at most the secondary sense of final causality, the "for which" as opposed to the "of which." The incidental ruining of one harvest by rain that comes just too late, or of another by a dry spell that comes too early, is in either case a result subordinate to the true and primary causes at work in the cosmos and in wheat.

When Aristotle argues that (*a*) nothing that happens for the most part is by chance and (*b*) everything that is not by chance is for the sake of something, the argument is not presented in a vacuum and meant to stand as a demonstration. It depends upon the recognition and definition of nature in II, 1. In that context, the argument shows that all events either are end-directed, by nature or by human choice, or result from the incidental combination of two or more activities that are end-directed. Chance does not fall outside the realm of ends, but depends upon ends. Final causes, causes for the sake of which, govern all other causes, incidental and non-incidental as well. It is clear that the material cause is subordinate to the others, and in nature the other three kinds of cause are one. The end of any natural being is to live the life given by its form, and the external source of its coming into being is the form transmitted by its parents into its embryonic material. The material has to be suitable, and hence it contributes necessary conditions of the natural being, and this is where Aristotle locates the realm of necessity. Stones and lumber don't stack themselves up by necessity into the form of a wall, but if there is to be a wall there must necessarily be something heavy and sturdy below and something lighter but stable above. Similarly, in a plant, roots do not go downward because they cannot help it, but because, given the needs of the plant, the growth downward of roots is conditionally necessary, necessary to serve the end which is the life of the whole plant. And the situation is the same whether the ends of the natural thing are met or missed. Unconditional necessity is no more possible than is unconditional chance. If you trip on a rock, you begin to fall down by necessity, but only because you are first of all a being with an upright posture. A worm would not trip. Everything in the world is subject to the workings of necessity and of chance, but only because it has an enduring identity given by its form. What happens by chance or necessity never stands alone as self-explanatory, but can only be understood in light of the ends things aim at by nature. All events and all motions are rooted in the potencies present by nature in things. This makes possible a comprehensive understanding of motion, which the next section of the *Physics* takes up.

Book III, Chapters 1–3

Motion

Chapter 1

200b, 12 Since nature is a source of motion and of change, and our pursuit is for nature, we must not let what motion is remain hidden. For it is necessary, being ignorant of it, to be ignorant also of nature. And once we have drawn a boundary around motion, we must try in the same way to advance upon the things that follow in succession.

Now motion seems to be one of the continuous things, and the infinite comes to sight first in the continuous; and for this reason, it often falls to those who define the continuous to need also an articulation of the infinite, on the grounds that the continuous is that which *20* is divisible infinitely. Besides these things, motion seems to be impossible without place and void and time. Both for these reasons, then, and because these things named are also generally common to all things, we must inquire by taking each of them in hand (for insight into particular things is later than that into those that are common). And first, as we said, we must take up motion.

There is that which is fully and actively itself, but also that which is what it is, in part, only potentially: either being a *this,* being this much, being of this kind, or similarly with the *30* other ways of attributing being. Being in relation to something is attributed to what exceeds or falls short, or to what acts and what is acted upon, or generally to what moves (something) and what is moved: for what moves is a mover of something moved, and what is moved is moved by something moving it, and there is no motion apart from things. For what changes always changes either in thinghood, or in amount, or in quality, or in place, and there is nothing to take hold of which is common to these, and is neither, in our manner of *201a* speaking, a *this,* nor a this much, nor an of-this-kind, nor any of the other kinds of being: so that neither motion nor change will be anything apart from the things named, since there is, in fact, nothing other than the things named. Now each of these may belong to anything in two ways: a this may be a form or its deprivation, an of-this-kind either (say) white or black, or a this-much either complete or incomplete. In the same way, then, also, a change of place may be either up or down, of something either light or heavy. Therefore, there are just so many kinds of motion and of change as there are of being.

10 A distinction having been made in each kind of being between the fully active and what *is* only potentially, the being-at-work-staying-itself of whatever is potentially, *just as such,* is motion: of the alterable, as alterable, it is alteration, of what can grow and its opposite, what can shrink (since no name is common to the two), it is growth and shrinkage, of the generable and destructible it is coming-to-be and passing away, and of the movable in place it is change of place.

That this is motion is clear from this: when the buildable, just insofar as it is said to be such, is fully at-work, [namely at-work-staying-buildable,] it is being built, and this is the activity of building. A similar formulation applies to the activities of learning, healing, *20* rolling, leaping, ripening, and aging. And since some of the same things *are* both potentially and in full activity, though not at the same time and in the same respect, but for example actively hot and potentially cold, many things will at once be both acting and being acted upon by each other, for all will at the same time be both active and passive. Thus what causes motion in a natural way is also moved, for each such thing moves both the moved and itself. To some, indeed, it seems that everything that causes motion is moved, but how this truly stands will be clear from other considerations (for there is something causing motion and motionless). But the being-at-work staying-itself of what *is* potentially, whenever, being fully at work, it is at work not as itself but just as movable, is motion. *30* By the "just as" I mean this. Bronze is potentially a statue, but it is not the being-at-work-staying-itself of bronze *as* bronze that is motion; for the being-bronze itself is not the being-potentially-something, since, if they were simply the same and *meant* the same thing, the being-at-work-staying itself of the bronze as bronze would be motion. But they are not the same, as was said. (This is clear in the case of contraries. For to be potentially healthy *201b* and to be potentially sick are different. If they were not, to be sick and to be healthy would be the same. But the subject underlying both the health and the sickness, whether blood or some other fluid, is the same and one.) Since then they are not the same, just as neither are a color and being-capable-of-being-seen the same, it is clear that the being-at-work-staying-itself of a potency, *as* a potency, is motion.

It is clear both that this is motion, and that a thing happens to be moved whenever this being-at-work-staying-itself is, and neither before nor after. For each thing admits at one time of being active, at another of not being active. An example is the buildable. The being at work of *10* the buildable, just as buildable, is building. (For the being-at-work must be either building or the house. But whenever the house is, the buildable is no longer. But it is the buildable that is being built. Necessarily then, building is the being-at-work.) But building is a certain motion. And surely the same account will exactly fit the other motions.

Chapter 2

That this has been stated well is confirmed both by the things others have said about it and by the fact that it is not easy to define it otherwise. For one could not even *20* place motion and change in any other class, though it is clear to those who consider [what has been written on the topic] how some place it, asserting motion to be otherness or inequality or non-being. But none of these is a necessary condition of being moved, namely that a thing be either other or unequal or not be, nor is change either into these or from these any more than into or from their opposites. But the reason for placing it in these classes is that motion seems to be something indefinite, while a whole array of negative principles seem also to be indefinite, since none of them is a *this* nor an of-this-kind nor belongs to any of the other ways of attributing being.

And the reason motion seems to be indefinite is that one cannot place it as a *30* potency of things or as a being-at-work. For neither the potency to be this-much nor the actively being this-much is necessarily a being-moved; so motion seems to be a certain being-at-work, but incomplete. The reason for this is that the potency, of which it is the [complete] being-at-work, is itself something incomplete. On account of this it is difficult to get hold of what it is. For one must place it as a deprivation, or a potency, or an *202a* unqualified being-at-work, but none of these seems admissable. There remains, then, the way stated, that it is a certain being-at-work, a being-at-work of such a kind as we have described, difficult to bring into focus, but possible to be.

Everything which causes motion is also moved, as has been said, if it is potentially movable and is something of which the motionlessness is rest. (For the motionlessness of that to which motion belongs is rest.) For to be at work upon the movable as such is to move it, and this a thing does by contact, so that at the same time it is also acted upon. Therefore, motion is the being-at-work-staying-itself of the movable, as movable, and happens to it by contact with what is moving, so that the latter too is acted upon. And what *10* moves will always bear a form, whether a *this* or an of-this-kind or a this much, which will be a source and cause of its motion whenever it moves. For example, a fully-at-work human being brings about another human being from what is potentially human.

Chapter 3

And the riddle° is now cleared up, since motion is in the thing moved. For the being-at-work-staying-itself (which the motion is) is *of* the thing moved, though from the thing causing motion. And the being-at-work of the thing causing motion is nothing other than this, for it must be the being-at-work-staying-itself of both. For a thing is capable of causing motion by its potency, but *is* moving by its being-at-work, and it is at work upon the thing moved. So the being-at-work of both is one, just as the interval from one to two *20* and from two to one is the same, and the uphill and downhill road. In these cases, the thing is one, though the meaning is not, and it is likewise with the causing motion and being moved.

But there is a logical impasse. For perhaps it is necessary that there be one being-at-work of the active and one of the passive, since the one is a doing and the other a being-done-to, and the work and end of one is a thing done, and of the other a thing suffered. Since, then, both are motions, if they are distinct, where are they? For if it is not the case that both are in the thing acted upon and moved, then the being-active is in the one acting and the being-passive in the one acted upon. (And if one must also call this latter "acting," it would be ambiguous.) But if this is so, the motion will be in the thing causing movement *30* (for the same description applies to moving and being moved as to acting and being acted upon), so that either everything that causes motion will be moved or a thing having motion will not be moved. But if both are in the thing moved and acted upon, both the acting and the being-passive, and both teaching and learning, though they are two, are in the learner, *202b* then, first, the being-at-work of each will not inhere in each, and further, it would be strange to be moved in two motions at the same time. What would be the two alterations of one thing changing into one form? It is impossible. So there will be one being-at-work. But it is unreasonable that there be one and the same being-at-work of two things different in kind. But this will be so if teaching and learning are the same [being-at-work], and acting and being-acted-upon, and to teach is the same as to learn, and to act as to be acted upon, so that the one teaching will have to learn everything, and the one acting be acted upon.

On the other hand, it is not absurd for the being-at-work of one thing to be in something else. (For teaching is a being-at-work of the one who can teach, but surely it is a being-at-work upon someone, and is not divided, but is *of* this person, *in* that one.) Nor does anything prevent one and the same thing from belonging to two things (though not in *10* the sense that the *being* is the same, but in the way in which what *is* potentially comes into being in relation to something already at work). Nor is it necessary that the one teaching be learning, not even if the acting and being acted upon are the same (again not in the sense that there is one articulation of what it means to be each, but in the manner of the road from Athens to Thebes and that from Thebes to Athens, the sort of example mentioned earlier). For things are not identical to which the same things belong in some particular way, but only those of which the being is the same. And it is by no means the case that to learn is the same as to teach, not even if the activity of teaching is the same as that of learning, just as the separation from here to there is not one and the same as that from there to here, *20* even though the interval between the things set apart is one. And to speak generally, teaching is not the same as learning in the highest and most proper sense, nor acting the same as being acted upon, but that to which these belong, the motion, is the same. For the being-at-work of this in that, and the being-at-work of this by the action of that, differ in meaning.

What motion is, then, has been said, both generally and for particular instances. For it is not unclear how each of the kinds of it will be defined: alteration, for example, is the being-at-work-staying-itself of the alterable as alterable. And still more explicitly, motion is the being-at-work-staying-itself of the potentially-active-or-acted-upon *as such*, both simply and in each case, such as building or healing. And this will be said in the same way about each of the other motions.

Commentary on Book III, Chapters 1–3

Nature, of which the entire inquiry of the *Physics* is in pursuit, was defined as an internal cause of motion and rest in that to which it belongs primarily. The things that have natures have been determined to be the animals and plants and the cosmos as a whole. The internal cause that gives each of them its nature has been shown to be form, understood as the being-at-work for the sake of which any of them does all that it does. Once motion has been defined, the definition of nature will have been fully unfolded. Aristotle is perhaps the only thinker who has ever attempted to define motion, rather than merely describing it or denying it. His definition has been called one of the great achievements of human reason, and it requires work to understand it.

For the past thousand years or so, almost everyone who has written about Aristotle's definition of motion has misunderstood it. The definition is constructed at the limits of thought and speech, and inadequate translation makes it crumble away to nothing. It did not travel well into Latin, and in the form in which it came into English from Latin it is scarcely intelligible. Aristotle says that there is only one class to which motion can be assigned: if motion is anything positive at all, rather than just a lack of something or departure from something, it must be some sort of *entelecheia*. The standard translation of this word is actuality, but even if there were no other reason to reject this translation, the nonsense it makes of the definition of motion would be reason enough. To call motion an actuality seems to say that it is not motion. Sir David Ross says that that is because *entelecheia*, in this one place only, means actualization. But to call motion an actualization says that it is a kind of motion. Short of going back to the translator's drawing board, there is no way to recapture the meaning of this definition.

Aristotle regards *entelecheia* as one of the ultimate terms of discourse, and not itself definable, but in Book IX of the *Metaphysics* (1047a, 30–31; 1050a, 21–23) he says that its meaning converges with that of *energeia*, and in IX, 6 he explains *energeia* by means of examples and analogies. The genus of which motion is a species is being-at-work-staying-itself, of which the only other species is thinghood. The being-at-work-staying-itself of a potency, as material, is thinghood. The being-at-work-staying-itself of a potency as a potency is motion. A thing, in the primary sense of the word, an independent thing, is a fusion of material, which *is* as potency, with the being-at-work that is its form. To be at all, a thing must be at-work-staying-itself, breathing and assimilating food if it is an animal, taking in nourishment from the earth and air if it is a plant, circling and maintaining the equilibrium of its parts if it is the cosmos. But an animal must be born, must grow, must develop, must go seek food; a plant must sprout, put forth roots, form leaves and seeds; and displaced pieces of the cosmos must travel up or down. These motions are all potencies staying-themselves as potencies, not fused into the states of active completion toward which they are potencies. That is what Aristotle says motion is.

But the definition seems to ignore all those motions that are not from nature. The two domains of causes other than nature, art and chance, both produce motions of their own. But art itself is a potency of human beings, by nature, so that the production of works of art is a natural motion. And the motions of artifacts, of the sands in an hourglass, for example, are nothing but natural motions harnessed and constrained by human skill. As for chance, we have seen that it is what arises when two lines of causes cross incidentally. Does a child have an inherent potency to be run over by a car? Surely not, in the strong sense of potency that Aristotle intends, but a child running into the street chasing a ball or a dog is giving way to an impulse inherent in his or her nature, and an adult driving home or to work or to the grocery store is also acting from some natural human yearning. Each motion fits perfectly into Aristotle's definition of motion, since only the intersection of the motions belongs to chance. And what of a human being falling off a cliff? That motion is a natural one for a chunk of earth, and a human being is incidentally partly earth. The accidental fall is not the expression of a human potency, but of a natural potency incidental to human beings. No special kind of motion beyond that in Aristotle's definition is at work. Within the comprehensive view of the world presented in Book II, no motion could exist that would be outside the definition.

The "riddle" mentioned at the beginning of Chapter 3 is Zeno's observation that a thing cannot be in motion either where it is or where it is not. The long shadow of Zeno extends from here over all the rest of the *Physics*. Zeno's paradoxes all aim at showing motion to be impossible or unintelligible, and they will turn up twice more in Book VI and again in Book VIII, one step from the end of the *Physics*. Zeno's arguments will never lose their freshness and power. They cannot be right, but there is no easy way to say how they are wrong. For Aristotle, the whole study of nature turns on understanding motion adequately. Where is a motion? To consider the motion as external to the moving thing makes this an insoluble puzzle, but Aristotle's definition, locating motion in the potencies of things, resolves it. A being-at-work of anything is where that thing is (unless it is at work upon something else, as a teacher is at work upon a student). For Aristotle, motion is not a brute fact of sensation, but is built into the very structure of being. Zeno's critique of motion is a *reductio ad absurdum* of the idea that there could be motion if there were not pre-existing potencies in things.

The remaining sections of Books III and IV take up in turn the conditions of the possibility of motion. That the first of these is the infinite is another glance in the direction of Zeno, since the most powerful of his arguments will exploit the infinity present in distances, times, and motions.

Book III, Chapters 4–8

The Infinite

Chapter 4

202b, 30 Since the knowledge of nature is about magnitude and motion and time, each of which must be either infinite or finite, even if not everything is infinite or finite, such as a feeling or a point (for there is perhaps no necessity that such things be among either of these), it would be an appropriate thing for the one concerned with nature to consider about *203a* the infinite, whether it is or not, and if it is, what it is. And a sign that the consideration about it is at home with this knowledge is this: all those who are regarded as having touched on such philosophy in a way worthy to speak of have made an account about the infinite, and all have set it down as a source of the things that are. Some, such as the Pythagoreans and Plato, set down the infinite by itself, not as incidental to anything else, but as itself an independent being. But the Pythagoreans place it among sensible things (for they do not make number separate from these), and make what is outside the heavens to be infinite, while Plato sets nothing outside body, not even the forms, on account of their not being in a place, but *10* still makes the infinite be present both in the sensible things and in the forms. And the former take the infinite to be the even. (For this, when it is taken into, and limited by, the odd, provides infinity to things. A sign of this is taken to be what happens with numbers, for when gnomons° are placed around the one and in other ways, sometimes the form comes out different each time, but sometimes it is a single one.) But Plato makes the infinites two, the great and the small.

But all those concerned about nature set down as underlying the infinite some other nature from among the so-called elements, such as water or air or what is between these. Of those who *20* make the elements limited in number, none makes them infinite [in extent]; but as many as make the elements infinitely many, which Anaxagoras and Democritus do, the former out of the homogeneous constituents, the latter out of the shapes of the mixtures of all seeds, say that the infinite is continuous by contact. Anaxagoras says that any of the parts is mixed in the same way as the whole, on account of seeing anything come to be out of anything; it is likely that this is his reason too for saying that all things were at some time together with their like: as with this flesh and this bone, so also with anything, and therefore with everything, and at the same time forsooth. For there is a source of dissolution not only in each thing but also of *30* all things. For since what comes into being comes from a body of such a kind, and of everything there is a coming-into-being, though not at the same time, and since there must be some source of the coming-into-being, and this is one such source, that which he calls intellect, and intellect works by thinking from some beginning, *therefore* it is necessary that all things were at one time together with their like and began at that time to be moved. But Democritus *203b* says that none of the primary beings comes about from another one; nevertheless, for him the common body is the source of all things, differing in its parts in size and shape.

That, then, this consideration is one that belongs to those who study nature, is clear from these things. And reasonably, too, all make it a source. For neither is it possible for it to be present with no effect, nor for any other power to belong to it except as a source. For everything either is a beginning or is from a beginning, but of the infinite there is no beginning, for it would be a limit of it. Further, it is both ungenerated and indestructible, as though it were *10* a source. For it is necessary both that everything that comes into being reach an end, and that there be a finishing off of every decay. For this reason, as we have said, there is no source of the infinite, but it seems to be a source of the other things, both containing everything and governing everything, just as all those say who make no other causes beside the infinite such as intelligence or friendship. And they make it the divine, for it is deathless and indestructible, just as Anaximander and most of the writers about nature say.

Of the being of something infinite, belief might come to those who consider the topic from five things most of all: from time (for this is infinite) and from the division of magnitudes (for the mathematicians too use the infinite); further, from there being only one way for coming into being and passing away not to give out, that is, if there were an infinite *20* from which what comes into being is taken; and again from the limited's always being limited by something, so that necessarily nothing could be a limit if each thing must always be limited by another. But greatest of all and most authoritative is what brings everyone to the familiar impasse: For since they do not give out in our thinking, both number and mathematical magnitudes seem to be infinite, as well as what is outside the heavens. But if what is outside is infinite, it seems that both body and worlds are infinite. For why, in the void, would they be in this place rather than in that place? So if in any place, then also in every, there would be bulk. So if void and place are infinite, body too must necessarily be *30* so at the same time; for among everlasting things, to be possible and to *be* do not differ in any way.

But the consideration of the infinite has an impasse, for many impossibilities meet both those who posit that it is not and those who posit that it is. Further, in what way is it? Is it in the manner of an independent being, or as incidental to and following necessarily *204a* from some other nature? Or is it in neither way, but is there nonetheless an infinite or an infinitely many? Most of all, though, it belongs to the one who studies nature to examine whether there is an infinite sensible magnitude. First, then, one must distinguish in how many ways the infinite is meant. In one way it is what is impossible to go through by not being of such a nature as to be gone through; in this sense sound is invisible. But in another way it is what has a way-out-through which cannot be completed at all, or scarcely so, or that which, being by nature such as to have a way-out-through or a limit, does not have one. Further, every infinite is so either by addition or by division or both.

Chapter 5

Now for the infinite to be separate from sensible things, itself being something *10* infinite, is not possible. For if it is neither a magnitude nor a multitude, but the infinite is itself an independent being and not an incidental one, it will be indivisible (for what is divisible will be either a magnitude or a multitude). But if it is indivisible, it is not infinite, except in the way that sound is invisible. But this is neither what those who say there is an infinite say it is, nor what we are inquiring about, which is what cannot be gone through completely. But if the infinite is something incidental, it could not be, as infinite, an element of things, just as the invisible is not an element of speech, even though sound is invisible. Further, how is it possible for the infinite itself to be something which number and magnitude are not, when the infinite is incidental to them and results from what they are? *20* For necessarily it will have any attribute less than the number or the magnitude has it.

It is clear too that it is not possible for the infinite to *be* either as something at-work or as an independent and original being. For whatever is taken from it will be infinite, if it has parts (for being-infinite and the infinite are the same, if the infinite is an independent being and not dependent on some underlying thing), so that it will be either indivisible or divisible into infinites. But for the same thing to be many infinites is impossible (but just as part of air is air, so also part of the infinite is infinite, if it is an independent and original being). Therefore it is without parts and indivisible. But this is impossible for what is infinite in the complete sense, for it is necessary that it be a quantity. Therefore the infinite *30* may be set down as something incidental. But if so, it has been said that it is not possible to call *it* an original source, but rather that to which it is incidental, the air or the even. So those who speak as the Pythagoreans do are exposed as speaking absurdly. For at the same time they make the infinite an independent being and divide it.

But perhaps this inquiry is whether, generally, it is possible for the infinite to be *204b* among mathematical things, or among intelligible things not having magnitude, while we are inquiring about sensible things, about which we are making our present pursuit, whether there is or is not among them a body infinite in extent. To those who consider the matter from a merely logical standpoint, in the following way, it would seem that there is none; for if the meaning of body is "that which is bounded by a surface," there could be no infinite body, either intelligible or sensible. (But surely in this way not even isolated number could be infinite. For number, *10* or what has number, is numerable. If, then, it is possible to count the numerable, it would also be possible to go through the infinite.)

Let us consider instead, in a way suited to the study of nature, the following: it cannot be either composite or simple. The infinite body will not be composite if the elements are limited in number. For they must be more than one, and the contrary ones must always be in equilibrium, so no one of them can be infinite. (For if the power in one body falls short to any degree whatever of that in the other, as if fire were limited but air infinite, while a quantity of fire is to an equal quantity of air in any ratio whatever in power, so long as it has some numerical amount, nevertheless it is clear that the infinite one will surpass and destroy the *20* finite one.) And it is impossible that each of them is infinite. For body is what has extension in every direction, and the infinite is what is extended limitlessly, so that the infinite body will be extended in every direction to infinity.

But neither is it possible that an infinite body be one and simple, neither, as some say, as what is apart from the elements, and out of which these come into being, nor in any way at all. There are some who make the infinite a separate body, and not air or water, lest the other things be destroyed by the infinite one among them. For they are adverse to one another: air is cold, for example, water moist, and fire hot. If one of them were infinite, the others would already have been destroyed. So now they say it is something else, out of which these things *30* come. But it is impossible that there be such a thing, not because it is infinite (for about this one must speak in common about everything in the same way, air and water and anything else), but because there *is* no such sensible body apart from the so-called elements. For everything is made of something, and dissolves into this, so that this infinite sensible body would be here alongside air and fire and earth and water; but there is no evidence of it.

205a And it is in no way possible that either fire or any other of the elements be infinite. For in general and apart from the being-infinite of one of them, it is impossible that the whole, even if it is finite, either be or become some one of them, as Heracleitus says that all things at some time become fire. (And the same argument applies to the one which the writers on nature make apart from the elements.) For all things change from contrary into contrary, as from hot into cold.

But in connection with each proposal, it is necessary to examine in the following way whether it is possible or impossible for it to be. And that, in general, it is impossible that *10* there be an infinite sensible body, is clear from these things. For it is the nature of every sensible thing to be somewhere, and there is some place for each, and it is the same for the part and for the whole, as for both the whole earth and one lump of earth, and for all fire and a spark. So first, if it is homogeneous, it will be either motionless or always carried along, and this is surely impossible. (For why downward rather than upward or any other way? I mean, if it were, say, a lump of earth, where would it be moved or where stay still? For the place of the body homogeneous with it is infinite. Then will it take up the whole place? And how? What, then, or where will the rest and motion of it be? Or is it at rest everywhere? Then it will not be moved. Or will it be moved everywhere? Then it does not stay still.) And this is the reason none of the writers on nature made the one infinite body fire or earth, but rather water or air or something between them, because it was plain that the place of each of the former is marked out, while the latter are ambiguously up and down.

20 But if the whole is heterogeneous, the places too will be heterogeneous. First, the body of the whole will not be one except by contact; further, the parts will be finite or infinite in kind. They cannot be finite (for some, such as fire or water, will be infinite in extent, and others not, if the whole is infinite, and such a one would be the destruction of its contraries). But if the parts are infinite and simple, and the places infinite, the elements too will be infinite; but if this is impossible, and the places are finite, so too will the *30* whole be finite. For the place and the body cannot fail to match each other. For neither is the whole place larger than the size the body can attain (and the body will no longer be infinite), nor is the body larger than the place, for there would be either a void or a body of such a nature as to be nowhere.

205b Anaxagoras speaks absurdly about the stillness of the infinite, for he says the infinite itself holds itself fixed. This is because it is in itself (for nothing else surrounds it), as though wherever something is, it is its nature to be there. But this is not true, for something could be somewhere by constraint and not by its nature. Then if the whole unquestionably is not moved (for what holds itself fixed and is in itself must necessarily be motionless), one must still say why it is not of such a nature as to be moved. It is insufficient to say so and leave *10* off. For it could even be that it is not moved because it has no other place to move to, but it does not follow that it is not in its nature. The earth also is not carried along, nor would it be if it were infinite, being bound by the center; but it would remain not because there were no other place where it could be carried, but because it was so by nature. Surely it would be possible to say that it holds itself fixed. If, then, for the earth this is not the reason—its being infinite—

but the earth is at the center because it has heaviness, and the heavy stays at the center, likewise also the infinite would remain in itself through some other cause and not because it is infinite and itself holds itself fixed.

20 At the same time it is clear that any part of it whatever needs to stand still. For as the infinite remains fixed in itself, so also whatever part is taken remains in itself. For the places of the whole and of the part are homogeneous, as is the downward place of the whole earth and a lump of earth or the upward place of all fire and a spark. So if the place of the infinite is in itself, of the part too it is the same. Therefore it stands still in itself.

In general it is obvious that it is impossible to say at the same time that body is infinite and that there is a certain place for bodies, if every sensible body has either heaviness or lightness and if the heavy has by nature a motion to the center and the light upward. For this would be necessary also for the infinite, but it is impossible either that the whole be affected in each way or the half in either; for how would you cut it in two? Or *30* how in the infinite will there be up and down, or extremity and middle? Yet every sensible body is in a place, and the kinds of place differ: up and down, before and behind, right and left. And these are not only relative to us and by convention, but are also marked out in the whole itself. But it is impossible that these be infinite. Simply, if it is impossible for *206a* there to be an infinite place, and every body is in a place, it is impossible that there be an infinite body. But surely what is somewhere is in a place, and what is in a place is somewhere. If, then, the infinite cannot even be a quantity—since it would be a certain quantity, such as two feet or three feet, since quantity means that—so also with what is in a place, because it is *some*where, and this is either up or down or in some other direction out of the six, but each of these is some kind of limit. That, then, there is no actually infinite body, is clear from these things.

Chapter 6

But that, if there is no infinite simply, many impossible things follow, is clear. *10* For there will be a beginning and an end of time, as well as magnitudes not divisible into magnitudes, and number will not be infinite. But whenever such a distinction has been made and neither way seems possible, there is a need for discrimination, and it is clear that in one way the infinite is, and in another way it is not. Now being is said of what *is* potentially or of what is in complete activity, and there is an infinite by addition or by division. And that there is no magnitude actually infinite has been said, but there is magnitude that is infinite by division; for it is not difficult to refute indivisible lines. What is left, then, is that the infinite is as potentiality. But it is necessary *20* not to take the being-potentially in the same way as if something were potentially a statue, since this will also *be* a statue, and thus there would also be an infinite which would be at-work. But since there are many ways of being, just as day is, or the athletic games which always come about one after the other, so also with the infinite. (For also with these things there is both being-potentially and being-at-work; for there are Olympic games both in the sense that the games are capable of happening and that they are happening.) And this is evident in different ways in time, in human beings, and in the division of magnitudes. In general the infinite is in this way: it is in what is taken always one after the other, while what is taken is *30* always finite, but always another and another. So being is meant in many ways, and the infinite must not be taken as a *this*, such as a man or a house, but in the way that day or the games are meant, to which being belongs not as to a thing, but in a constant coming into being and passing away, finite, but always other and other. But in magnitudes what is taken *206b* remains, while with time and human beings it is always perishing in such a way as not to run out.

But the infinite by addition is in some way the same as that by division, for the latter comes about in the finite by addition turned back the other way. For where a division is seen to be to infinity, there is obviously an addition to what is cut off. For if someone taking a marked-off part of a finite magnitude keeps taking from it in the same ratio (not including the piece of the whole magnitude already taken), the pieces will not exhaust the finite thing. *10* But if in the same way one increases the ratio so as always to include the same amount, they do exhaust it, through the whole finite thing's being used up by whatever part is marked off. So it is in no other way, but in this way there is an infinite, in potentiality and by exhaustion (but it is also at-work, in the way we say day and the games to be). And it *is* thus in the way material is, potentially, and not on its own in the way the finite is. And the infinite by addition is surely in potentiality in the same way, which we say is in a certain way the same as that by division. For there will always be something outside to take, and it will not exceed *20* every magnitude, just as in the division it does go beyond every marked-off piece and there will always be a smaller piece.

Therefore to exceed everything by addition is not even possible potentially, unless there is accidentally an actual infinite, as the writers on nature say that which is outside the body of the cosmos, being of air or some other such thing, is infinite. But if it is impossible for there to be an actually infinite sensible body in this way, it is clear that not even potentially could there be one by addition, other than in the way described, by a reversed division. Plato too for this reason made the infinites two, because both in increase and in *30* reduction it seems to go on and on to infinity. Yet though he made them two he does not use them. For among the numbers belongs neither the infinite by reduction (since the unit is the smallest), nor that by increase (for he makes an [eidetic] number° only as far as ten).

The infinite turns out to be the opposite of the way people speak of it. For this is *207a* the infinite: not that outside of which there is nothing, but that outside of which there is always something. Here is a sign: people speak of rings which do not have stone-settings as endless because there is always something beyond to take, speaking in accordance with a likeness though not strictly. For it is necessary both that this condition be present and that at no time the same part be taken; but in the circle it does not happen that way, but only the succeeding part is always different. Infinite, then, is that of which, to those taking it by quantities, there is always something beyond to take. That of which nothing is outside is *10* complete and whole; that is how we define the whole, as that of which nothing is absent, as a whole human being or box. And just as with each example, so also in the strict sense, the whole is that outside of which there is nothing; but that of which something absent is outside is not entire, whatever might be absent. But whole and complete are either completely identical or closely akin in nature. And nothing is complete which does not have an end, and an end is a limit.

For this reason one ought to regard Parmenides as having spoken better than Melissus. For the latter calls the infinite a whole, but the former says the whole is finite, "equally-poised from the center."

For it is not joining a thread to a thread to connect the infinite with the *20* whole or the sum of all things, yet it is from this that they derive the grandeur attaching to the infinite, the containing all things and having all in itself, through its having some likeness to the whole. For it is true that the infinite is the material of the totality of magnitude, and is what is potentially whole, but not actually; and it is divisible not only in respect to reduction but also in the addition which is its reversal; and it is even whole and limited, though not by itself but by something else; and it does not contain but is contained, insofar as it is infinite. Hence also it is unknowable, insofar as it is infinite, for material does not have form. So it is clear that the infinite is within the meaning of part rather than that of whole. For the material is part of the whole, as the bronze of a bronze statue. And if it were to contain in the realm *30* of sensible things, it would have to be the case also in the intelligible realm that the great and the small contained the intelligible things. But it is absurd and impossible for the unknowable and indeterminate to contain and define.

Chapter 7

It is in agreement with reason to think there is no infinite by addition of such a kind as to exceed every magnitude, but by division there is. (For material and the infinite *207b* are contained within, while the form contains.) It is good reasoning too to think that in number there is a limit on the side of the least but on the side of the greater a number always exceeds every multitude, while magnitudes, on the contrary, go beyond every magnitude on the side of the less but on that of the greater there is no infinite magnitude. The cause is that the one is indivisible, whatever might be one (as a human being is one human being and *10* not many), but a number is many ones and a certain amount, so it is necessary to stop at the indivisible (for three and two are surnames, [numbers *of* something,] and likewise each of the other numbers), but it is always possible to think of a greater, for the bisections of a magnitude are infinite. So there is an infinite potentially, but not actually; but what is taken always goes beyond every determinate amount. But the infinite is not a separate number, nor is infinity static, but it is becoming, as are time and the number of time. And with magnitudes it is opposite, for the continuous is divided into infinitely many parts, but on the side of the greater it is not infinite. For however much it belongs to it to be potentially, also actually it is possible for it to be this much. So since there is no infinite sensible magnitude, it is not possible that *20* there be an excess beyond every determinate magnitude, for there would be something greater than the heavens.

The infinite is not the same in a magnitude and a motion and a time, as though it were some single nature, but what is derivative is spoken of in accordance with what is prior; for example, a motion [is spoken of in a certain way] because of the magnitude over which it moves or alters or increases, but a time on account of the motion. We use these notions now, but later we will also say what each is, and why every magnitude is divisible into magnitudes.

This account does not deprive the mathematicians of their study, though it does do away with anything's being infinite in such a way as to be actually untraversable in the direction of increase. *30* For as it is, they have no need of the infinite (for they do not use it), but they need only that something finite can be as great as they want. And it is possible that another magnitude of any size whatever be cut out in the same ratio as the greatest magnitude. So for the sake of demonstrating, it will make no difference to them whether the infinite is among the magnitudes there are.

And since the causes have been distinguished in a fourfold way, it is clear that the *208a* infinite is a cause as material, and that the being of it is a negation, while the underlying thing to which it belongs is what is continuous and sensible in its own right. And it is obvious that everyone else makes use of the infinite as material, for which reason it is absurd to make it what contains but not what is contained.

Chapter 8

What is left is to go through the arguments according to which the infinite seems to be not only potentially but as something determinate. Some of these have no necessity, while the rest have other things coming up against them that are true. For it is not necessary, in order that coming into being not give out, that there be an actually infinite *10* sensible body; for it is possible that the destruction of one thing be the coming into being of another, the whole being finite. Further, touching and being limited are different. For the one is relative and of something (for every touching is a touching of something), and is incidental to something among the limited things, but the limited thing is not relative to anything, nor is there touching of any chance thing by any other [see 203b, 20–22]. And to place trust in thinking is absurd, for the exceeding and the falling short are not in the thing but in the thinking. For someone might think each of us many times as big as he is, expanding him to infinity. But this does not make anyone bigger than a city or than whatever size we have, just the fact that someone thinks it, but only the fact he is so; this thinking is incidental. Time and *20* motion are infinite, and thinking as well, but what is taken does not persist. No magnitude is [actually] infinite, either by reduction or by increase in thought. In other respects about the infinite, in what way it is and is not and what it is have been said.

Commentary on Book III, Chapters 4–8

Many of Aristotle's materialist predecessors spoke of the world, or some kind of body in it, as infinite. The Pythagoreans, who said everything is number, saw each thing as a combination of the natures of the odd and even, the former containing things within limits, the latter being the indeterminate, limitless content. They found an image of these two natures in a certain way of displaying numbers as geometrical shapes. The ancient Greek astronomers had used the word gnomon to mean a stick stuck upright in the ground, by which the position of the sun could be known by the length and direction of the shadow. From the shape of the stick plus its shadow, the word came to refer to the shape of a capital L. Let three dots be arranged as an equal-armed *gnomon,* with one at the corner and one at each end. Placing this gnomon around one dot at the open corner gives the shape of a square, made from the first square number of dots, four. Since the unit and the first odd number made the first square number, try the next odd number, five, as a gnomon, giving two sides of a three-by-three square. Sure enough, it fits around the four-dot square to form the same shape again, with nine dots. And the series of odd numbers will continue in this way to give always and only the square figure as its sums, picking out all the square numbers. Even numbers of dots, on the other hand, can never be arranged in equal-armed gnomons; four dots, for example, give only the sides of a three-by-two rectangle, and when this is fit around two dots, the two-by-one rectangle, the shape changes, and it will continue to change every time the gnomon of the next even number is added.

Plato, in the technical side of his thinking, which does not appear in the dialogues, reduced all things to the one, as source of limit, identity, and being, and the indeterminate dyad, or great and small infinite, as that on which limit is placed. The first results of this combination are the most encompassing of the forms, which have the structures of numbers, with the one as the good, the determinate dyad as being, the soul as the eidetic (qualitative) three, virtue the eidetic four, and so on. When Aristotle says that Plato made numbers only up to ten (206b, 32), he is referring to these assemblies of forms. A discussion of these school doctrines of Plato can be found in Jacob Klein's book *Greek Mathematical Thought and the Origin of Algebra* (Cambridge, Mass.: M.I.T. Press, 1968, chap. 7, part C). For both the Pythagoreans and Plato, the infinite exists only as already tamed and bounded within some sort of mathematical limits.

Those who study nature, though, claim to find within it something unlimited. Aristotle argues, first, that this could not be simply "the infinite," standing on its own, but must be something that has infinity among its attributes. But is this one of the elements, or something else? It cannot be an element, since the elements are present as pairs of contraries, holding the powers of each other in equilibrium, and there is no evidence of any bodily nature in the world other than the bodily natures of the elements. Chiefly, though, Aristotle rests his conclusion that there can be no infinite body on the fact that this would destroy the conception of the cosmos as an ordered whole, with regions that are the places of each kind of thing. This argument, then, is provisional as it stands here, depending on the analysis of place in the next section, and finally on that of the void in the section after that. This is not the order of synthetic presentation, but the analytic order of inquiry.

But every magnitude, geometrical or bodily, is subject to infinite division, and within mathematics, magnitude and number can be infinitely increased. But these infinities are like the duration of the

Olympic games, which is only a few days each time they occur, but infinite as a future possibility. *Day* exists in the same way, twenty-four hours at a time, but in an ongoing sequence that has no end. This is what the infinite always is, according to Aristotle: a potential for increase beyond the present boundary or for division into ever more parts. No infinite is ever a whole or a *this*. As a kind of potency, the infinite has the nature of material, but it is a lesser kind of potency than the potencies that make natural beings come to possess form. As the capacity for one more finite boundary or division following any actual boundary or division, the infinite is the material of quantity, mere extensiveness or numerousness as such. And even this does not belong to natural beings in the direction of increase; people can be bigger than cities, and the cosmos bigger than any limit, only in our imaginations. The famous argument for the infinity of the world given by Lucretius in *De Rerum Natura* (I, 968–983), based on imagining throwing a javelin outward from whatever is purported to be the edge of the world, is just such a confusion of the way imagination works with the way things are.

Book IV, Chapters 1–5

Place

Chapter 1

208a, 27 In the same way the one who studies nature must also know about place, just as about the infinite, whether it is or not, and in what way it is, and what it is. For everyone assumes that beings are somewhere (and that what is not is nowhere; for where is a goat-stag or a sphinx?), and of motion the most common kind and the primary sense of the word is change of place, or what is called locomotion. But what place *is* holds many impasses, for it does not look the same to those who consider everything that belongs to it. And yet we *208b* have nothing about it from anyone else, pointing the way either to or through those impasses.

That place is seems to be clear from mutual replacement. For where water is now, when it has gone out from there as from a pitcher, air in turn is present, and at some time some other of the bodies occupies this same place, and this place seems to be different from all the things becoming present and exchanging with one another. For in that in which air is now, water was before, so that it is clear that there was a certain place or region different from both, into which and out of which they changed about. And further, the changes of place of the simple natural bodies, such as fire and earth and things of that kind, give *10* evidence not only that place is something, but also that it has some power. For each of them is carried into its own place if it is not obstructed, the one up and the other down; but these are parts or kinds of place, the up and the down and the rest of the six directions. And there are such things, the up and down and right and left, not only in relation to us. For with us they are not always the same, but follow our position, however we might turn (for which reason the same thing is often right and left, up and down, before and behind), but in nature each is marked off apart. For up is not whatever it happens to be, but is *20* where fire and what is light are carried; and likewise down is not whatever it happens to be but is where things having heaviness and earthen things are, as though they differed not only in position but in power. And mathematical things also give evidence, for though they are not in a place they still have a right and left according to situation relative to us, as things meaning only situation, not having each of them by nature. Further, those who say there is a void mean a place, for the void would be a place deprived of body.

That, then, place is something besides bodies, and that every body is in a place, one might accept from these things. And Hesiod would seem to speak rightly in making chaos *30* be first. At any rate, he says, "First of all things chaos came into being, but then broad-breasted earth," as though there had first to be present room for beings, through believing, as most people do, that every thing is somewhere and in a place. But if this is so, the power *209a* of place would be something wonderful, and pre-eminent among all things. For that without which none of the other things is, but which is without the other things, must necessarily be primary; for place is not destroyed when the things in it are annihilated.

It is not without an impasse, though, if place has being: what it is, whether some sort of bulk of body or some other nature. For one must first seek out its genus. It has, then, three extensions, length, breadth, and depth, by which every body is bounded. But it is impossible that place *be* body, for two bodies would be in the same place. Further, if body has a place or region, *10* it is clear that so do a surface and the rest of the boundaries, for the same argument will fit. For where the surface of the water was before, there in turn will be that of the air. But surely we have no difference at all between a point and the place of a point, so that if its place is not different from it, neither will the places of any of the other boundaries be, and place will not be anything aside from each of them. What then shall we set down place as being? For something having such a nature could not be either an element or made of elements, either of bodily or bodiless ones. For it has magnitude, but is not a body. But the elements of sensible things are bodies, and out of intelligible things no magnitude comes about. And further, of what among the beings could one set down place as the cause? For none of the *20* four causes is present in it, not as material of things (for nothing is composed of it), nor as form or articulation of things, nor as an end, nor does it move things. And further, if it is itself any of the beings, it will be somewhere. For Zeno's impasse seeks some explanation: if every being is in a place, it is clear that there will also be a place of the place, and so to infinity. Further, just as every body is in a place, so too in every place there is a body; how then are we to speak about growing things? For it is necessary from these things that the place grow along with them, if the place of each is neither smaller nor larger than it. *30* Through these considerations, then, it is necessary to be at an impasse not only about what it is, but even about whether it is.

Chapter 2

Now since something is meant either in virtue of itself or in virtue of something else, and a place is either the common one in which all bodies are or the private one in which something primarily is (I mean, for example, you are now in the heaven because you are in the air while it is in the heaven, but you are also in the air because you are on the earth, but *209b* also similarly you are on the earth because you are in this spot which contains nothing more than you); then if place is what primarily contains each of the bodies, it would be a certain limit, and so place would seem to be the form or shape of each thing, by which its magnitude, or the material of its magnitude, is bounded, since this is the limit of each one. To those who consider place in this way, the place of each thing is form, but insofar as place seems to be the extension of the magnitude, it is material, for the extension is different from the magnitude: it is that which is contained by the form and bounded as by a surface or limit, and this is the sort of thing that material and the indeterminate are. For when the boundary and attributes of the *10* sphere are taken away from it, there is left nothing aside from the material. And for this reason Plato says in the *Timaeus* that material and extension are the same, for that which participates [in a form] and extension are one and the same. Though he spoke in different ways there and in the things he said as unwritten teachings about that which participates, still he said plainly that place and extension are the same. For while everyone says that place is something, he alone made the effort to say what it is.

It is reasonable from these things, then, that it would seem to be difficult for those considering the subject to know what place is, if it is one of these two things at all, the *20* material or the form; in other respects, these involve the utmost height of examination, and it is not easy to tell them apart from each other. But in fact, it is not difficult to see that it is impossible for place to be either of these. For the form and the material are not separated from the thing, but the place can be. For in that in which air was, in this in turn water comes to be, as we said, with the water and air and other bodies as well exchanging with one another, so that the place of each is neither a part nor a condition of it but is separate. And place seems to be something of the kind that a jar is (for the jar is *30* a portable place), but the jar in no way belongs to the thing. To the extent then that it is separate from the thing, it is not the form, but to the extent that it contains, it is different from the material. And it seems that always that which is somewhere is itself something but something else is outside it. (And surely Plato ought to have said, if one may digress, why the forms and the numbers are not in place, if what *210a* participates is place, whether this in turn is the great and the small or, as he has written in the *Timaeus,* material.)

Further, how could something be carried to its own place if place were material or form? For it is impossible that anything that has no motion, nor an up or down, be place. So it is among such things that place must be sought. But if a thing's place is in it (as it must be if it is form or material), place will be in a place. For the form and the indeterminate change and move at the same time as the thing, and they are not always in the same place but are where the thing is; so there would be a place of the place. *10* Further, whenever water comes into being out of air, the place would be destroyed, since the body that comes into being is not in the same place. But what sort of destruction is this?

Why it is necessary that place be something, and in turn why one might be at an impasse about the being of it, have been said.

Chapter 3

After these things, one must take up in how many ways one thing is said to be *in* another. In one way, it is as the finger is in the hand, and generally the part in the whole. But in another way it is as the whole is in the parts; for the whole is not outside the parts. And in another way it is as human-ness is in animal and generally the species is in the genus. *20* But in another it is as the genus is in the species and generally the part of the species in its definition. Further, it is as health is in what is hot and cold and generally the form in the material. Again, it is as the affairs of the Greeks are in the king and generally in the first mover. Further, it is as a thing is in the good and generally in the end, and the end is that for the sake of which. But of all the ways, the most authoritative is that in which something is in a container and generally in a place.

One might be at an impasse whether it is also possible for a thing itself to be in itself, or not, so that everything is either nowhere at all or in something else. But it is possible to take this in two ways, considering the thing either in virtue of itself or in virtue of something else. For whenever they are parts of a whole, both that in which and also that which is in it, the whole will be said to be in itself. For a thing is also spoken of in reference to its *30* parts, as for example white, because its surface is white, or knowledgeable because of its reasoning part. So while the urn will not be in itself, nor will the wine, the urn of wine will be, for both that which is in and that in which it is are parts of the same thing.

In this way, then, it is possible for a thing itself to be in itself, but in the primary sense it is not possible. For example, the white is in the body (for the surface is in the body), but *210b* the knowledge is in the soul, and it is by way of these things, which are parts, that those names are in the human being. (But the urn and the wine are not parts when they are separate, but when they are together; hence, when they are parts, a thing itself will be in itself.) So the white is in the human being because it is in the body, and in the body because it is in the surface, but in this no longer on account of anything else. And these are different in kind, and each has a different nature and power, namely the surface and the white. And so with those who look at the subject by examples, we do not see anything in itself in any *10* of the meanings distinguished, and by reasoning it is clear that it is impossible. For each of two things would have to be present as both, the urn, say, as both jar and wine or the wine as both wine and urn, if it is possible that a thing itself be in itself. So that if as much as possible they were in one another, the urn would receive the wine not insofar as it is itself wine but insofar as that is, and the wine would be present in the urn not insofar as *it* is itself an urn but insofar as *that* is. That they are different in respect to the being of them is clear, for the meaning of that in which something is, is distinct from that which is in it. And surely neither is it possible incidentally, for two things will be in the same place at the same time. *20* For the urn itself would be in itself, if that of which the nature is receptive can be in itself, and moreover that of which it is receptive, the wine, say, if it is wine, would be in it.

That, then, it is impossible that a thing be in itself primarily is clear. And what Zeno was at an impasse about, that if place is anything it will be in something, is not difficult to resolve. For nothing prevents the primary place from being in something else, though not in that as in a place, but as health is in what is hot as a state, and the hot is in the body as an attribute. Therefore it is not necessary to go to infinity. And that point is clear, that since the jar in no way belongs to what is in it (for that primarily *in* and that *in which* are *30* different), place could be neither material nor form, but is something else. For both of these, the material and the form, belong to that which is present within. Let these, then, be the things raised as impasses.

Chapter 4

Now what place is might become clear in this way. Let us take as many things as seem truly to belong to it in its own right. We hold place to be primarily that which *211a* surrounds that of which it is the place, and in no way belongs to the thing; further, that the primary place is neither less nor greater than the thing; that it is left behind by each thing and is separate; and in addition to these things, we hold that all things having place have the up and the down, and each of the bodies is carried by nature to and remains in its proper place, which makes it be either up or down. Having laid down these things, one must examine the remaining ones. And it is necessary to attempt the investigation in such a way as to make what it is be delivered up, both so that the impasses be resolved and so that the things that *10* seem to belong to place will belong to it, and further so that the cause of the headache and the impasses about place be made clear. Thus each thing would be brought to light in the most beautiful way.

First, then, it is necessary to understand that place would not have been inquired about if there were not motion with respect to place. For on this account we regard the heaven most of all as in a place, because it is always in motion. And of this kind of motion there is not only change of place but also increase and decrease; for in increase and decrease too a thing changes, and what was in one place before is removed to a smaller or larger one. And a thing is moved either in its own right by being at work, or incidentally. *20* And of what is moved incidentally there is that which admits also of being moved in its own right, such as the parts of a body or the nail in a ship, but also things which do not so admit, such as whiteness and knowledge; for these have changed place because that in which they belong changes.

When we say that we are in the heavens as in a place, it is because we are in the air and this is in the heavens; and we are in the air but not in all of it, but on account of the innermost surrounding part of it we say we are in the air. (For if all the air were a place, each thing and its place would not be equal, but they seem to be equal, and that in which *30* something is primarily is of this sort.) Then whenever what surrounds is not divided but is continuous, a thing is said to be in it not as in a place but as a part in a whole; but whenever it is divided and touching, the thing is first of all in the innermost part of what surrounds it, which is neither part of the thing in it nor greater than its extension, but equal to it, for the extremities of things which touch coincide. If the surrounding thing is continuous, a thing *211b* is moved not in it but with it, but if the surrounding thing is divided, a thing is moved in it, and no less so whether the surrounding thing is moved or not. Examples of things not divided, but spoken of as parts in wholes, are the eyeball in the eye and the hand in the body, and of things divided the water in the jar and the wine in the jug, for the hand is moved with the body, but the water is moved in the jar.

Already, then, it is clear from these things what place is. For there are just four things of which it is necessary that place be some one: the form, the material, some sort of extension between the extremities, or the extremities if there is no extension besides the *10* magnitude of the body present within. But that it is not possible for it to be three of these, is clear. It is because it is a surrounding thing that it seems to be the form, for the extremities of what surrounds and of what is surrounded coincide. They are both boundaries, but not of the same thing, the form being the boundary of the thing and the place being that of the surrounding body. And it is because of the frequent changing of what is surrounded and separate, such as water from a jar, while what surrounds remains, that what is between seems to be some sort of extension, as though it were something apart from the body replaced. But there is no such thing; rather, some chance body falls in from among bodies of such a nature as to be adjacent and change places. But if there were some extension of such *20* a nature as also to remain, there would be infinitely many places in the same thing (for when the water and air change places, all their parts will be doing the same thing in the whole as all the water is doing in the jar). But at the same time the place will be changing, so that there will be also another place *of* the place, and many places will be together. And there is not another place of the part, in which it is moved, when the whole jar changes place, but it is the same; for the air and the water, or the parts of the water, replace each other where they are, not in the place in which they are coming to be, which is part of the place which is the *30* place of the whole heaven. But also the material might seem to be place, at least if one were to consider it in something at rest and not separated but continuous. For just as if something alters, there is something which now is white but formerly was black, or now is hard but formerly was soft (which is why we say that the material is something), so also on account of some such imagining place seems to be material, except the material is something because what was air, this now is water, but place seems to be material because where air was, there now is water. But material, just *212a* as was said above, neither is separate from the thing nor surrounds it, but place is both.

If, then, place is none of the three, neither form nor material nor some extension always present, different and apart from that of the thing displaced, it is necessary that place be the remaining one of the four, the boundary of the surrounding body at which it borders on the surrounded one. I mean by the surrounded body that which is movable with respect to place. But place seems to be something great and difficult to get hold of, both because *10* the material and the form are reflected alongside it, and because the removal of what is carried off happens in a surrounding body that is at rest; for that seems to allow there to be some extension within, other than the magnitudes moved. And the air too contributes something, because it seems to be bodiless, for the place seems to be not only the boundaries of the jar but also what is between them, as though a void. But just as the jar is a portable place, so also place is an immovable jar. For this reason, whenever in a moving thing, something inside moves and changes, such as a boat in a river, it is used in the manner of a jar rather than in that of a surrounding place. But place is meant to be motionless, on account of *20* which it is rather the whole river that is a place, since the whole is motionless. Therefore, this is place, the first motionless boundary of what surrounds.

For this reason, the center of the heavens and the surface toward us of what moves in a circle seem to be up and down in the most authoritative sense for everyone, because the one always holds still, while the surface of the one in circular motion remains holding itself in the same position. So since what is light is carried up by nature, and what is heavy is carried down, the surrounding boundary up against the center, and the center itself, are down, and that up against the extremity, and the extremity itself, are up; and for this reason place and what surrounds seem to be a certain surface, and like a jar. Further, the place coincides *30* with the thing, for the boundaries coincide with the bounded.

Chapter 5

That body which has some body outside surrounding it is in a place, but that which has not is not. But even if water were to come to be uncontained, the parts of it would be moved [with respect to place] (for they are contained by one another), but in one way the whole would be moved and in another way it would not. For as a whole it would at the same time not change its place but be moved in a circle, and this would be *212b* the place of its parts, some of which would move not up and down but in a circle; but some parts would move up and down, as many had the capacity to become dense or rare.

As was said, some things are in a place potentially, others actually. So whenever something homogeneous is continuous, the parts are in places potentially, but whenever it is divided up and touching, like a heap, they are actually so. And some things are in places in their own right. (For example, every body that is movable either by change of place or by increase is somewhere in its own right, but the heaven, as was said, is not anywhere, nor *10* is it as a whole in any place, if no body surrounds it; but on that upon which the heaven is moved there is a place for its parts, since one of the parts is bordering on another.) But other things are in places incidentally, such as the soul and again the heaven, since all its parts are in places in some way, because on the circle one part flanks another. For this reason the upper part is moved circularly, while the whole is nowhere. For what is somewhere both is something itself and needs there to be something else besides it, in which it is and which surrounds it. But aside from the whole or the sum of all things, nothing is outside the sum of all, and for this reason all things are in the heaven, for the heaven is equally the whole. But it is not the *20* case that the heaven is place; rather, some boundary of it is the place of the movable body it is touching. And in this way, the earth is in the water, and the water is in the air, and the air is in the aether, and the aether is in the heaven, but the heaven is no longer in anything else.

And it is clear from these things that all the impasses could also be resolved when place is articulated in this way. For the place need not grow along with a thing, nor a point have a place, nor two bodies be in the same place, nor need there be some sort of bodily extension (for what is within the place is the body that happens to be there, but not an extension of body). And place too is somewhere, but not as in a place, but rather as the limit is in what is limited. For not every being is in a place, but only movable bodies. And *30* it is reasonable that each thing is carried to its own place. (For what is next to something and touching it, not by constraint, is of the same kind, and things of the same nature are unaffected, while those which interact become passive to and active upon one another.) And everything stays by nature in its proper place, not unreasonably. For a part does, and what is in a place is like a separated part in relation to a whole, as when one moves a part *213a* of water or of air.

Further, air has to water this relation: that of form to material. For the water is material for the air, and the air like a certain being-at-work of it, since water is potentially air, though air is potentially water in a different way. One must make distinctions about these things later; but for this occasion it was necessary to mention them, and what is said unclearly now will be clearer then. If, then, the material and its being-at-work-staying-itself are the same (for water is both, the one potentially and the other in full activity), the water would be in relation to the air in some way as a part in relation to a whole. For this *10* reason there is contact between them, but they are of the same nature whenever they become one actively.

About place, both that it is and what it is have been said.

Commentary on Book IV, Chapters 1–5

Perhaps the most surprising thing about Aristotle's *Physics* to a present-day reader is the absence of the concept of space. We are so accustomed to thinking of all events as taking place in some vast empty expanse that we see no way to look at the world without it. It is not that this conception did not occur to Aristotle. At 211b, 16–19, and again at 212b, 25–27, he describes the idea of space and rejects it. We imagine the extension of a body, take away the body itself, and think that what is left in our imaginations is something present in the world. This is exactly the error described at the end of Book III. When Lucretius thinks he has proven the infinity of the world, he has in fact only proven the infinity of space, an object of the geometrical imagination. In the Scholium to the Definitions of his *Principia,* Newton calls this "absolute, true, and mathematical space," a homogeneous medium which bears no relation to sensible bodies. It is in this medium that the demonstrations of Euclid or any other geometer take place, but to confuse it with the world we live in is a very different affair.

For Aristotle, the evidence that there is such a thing as place comes not from trying to imagine directly what everything is in, but from thinking about our experience. Not only does one thing come to be where another thing was before, but where things are seems to matter to them. All things seem to have natural places where they can be at rest; in places other than their own regions of the cosmos, things are dis-placed, held there only by constraint, and they return spontaneously to their proper places when all obstacles are removed. Unlike Newtonian space, places are heterogeneous. Trees do not take root in air, human beings do not breathe in the sea, and stars do not circle underground. A thing that changes place is going toward, being removed from, or moving within a place in which it can remain and sustain itself, and this depends entirely on the kind of thing it is. Place, then, is an idea that presupposes an organized cosmos. We and all other beings have not just relative positions, but lives and activities that can only take place in an appropriate environment, which the cosmos not only makes room for, but sustains and nourishes.

The first stabs at understanding place that Aristotle describes are the three-dimensional outline of a thing, the crudest notion of form, or the expanse contained within that outline, the emptiest notion of material. But the truer conception of material as potency spilling over into form, and of form as the being-at-work of material, makes these clearly inseparable from one another or from the thing, while its place is something from which it can be removed. A third possibility is the one considered above, that of space as extension without body, which Aristotle considers a mere misuse of our own mathematical imaginations. He concludes that place is the inner boundary of the outer surrounding body—not of any chance surrounding body but of some stable component of the cosmos. Those things have places that have a home within the cosmos, while the whole of the cosmos has no place at all; it is literally nowhere, but is self-contained.

This account of place leaves one thing unexplored. Could there be one or more finite regions within the cosmos that are void of any body? That becomes the topic of the next section.

Book IV, Chapters 6–9

The Void

Chapter 6

213a, 12 In the same way, the one who studies nature must take it in hand to consider also about the void, whether it is or not, and in what way it is, and what it is, just as about place. For disbelief and belief about it are much the same as about place, according to what is taken up. For example, those who speak of the void set it down as a sort of place or container, and it seems to be full whenever it holds the bulk of which it is receptive, but void whenever it is deprived of it, as though what is void and full and a place were the same thing, *20* though the being of them is not the same. But it is necessary to begin the inquiry by taking up what is said by those who say there is void, and in turn what is said by those who deny it, and third the common opinions about what is said.

Those who try to show that there is no void do not refute what people intend void to mean, but what they themselves say in a mistaken way; Anaxagoras and those who refute in the same way do just this. For they prove that the air is something by twisting wine-skins and showing how strong the air is, and by drawing off liquids in clepsydras.° But people mean the void *30* to be extension in which there is no sensible body; and supposing everything that *is* to be body, they say that that in which there is nothing at all is void, for which reason that which is full of air is void. It is not therefore this that needs to be shown, that the air is something, but that there is no extension different from that of bodies, either separable or actually separate so as to take apart the whole body, so that it is not continuous, as Democritus and *213b* Leucippus and many other writers about nature say, or even whether there is something outside the whole of body, if it is continuous.

These people, then, do not even approach the doorway in front of the problem, but rather those do who say there is a void. And one thing they say is that there would be no motion with respect to place (that is, change of place or increase), for it does not seem that there would be motion if there were no void, since it would be impossible for what is full to receive anything. But if it did receive, and there were two things in the same place, then it would also be possible for there to be any number of bodies whatever together, since one could *10* tell no difference, through which it would be possible to deny it. But if this is possible, then also the smallest place could receive the largest thing, for the large one is many small ones. Melissus even demonstrates from these things that the whole is motionless, for if it were to be moved, a void would be necessary (he says), but the void is not among the things that are.

In one way, then, they demonstrate from these things that there is some sort of void, but in another way they demonstrate it because it is obvious that some things contract and are compressed, in the way that they say the vats hold the wine along with the wineskins, as though the close-packed body *20* were contracted into the voids present within. And further, growth seems to happen in everything by means of void, for the nourishment is a body, and it is impossible that two bodies be in the same place. Also, they make evidence out of what happens with ashes, which accept equally as much water as does the empty jar.

The Pythagoreans too said there is a void, and that it comes into the heaven from the infinite breath, as though the heaven were inhaling the void, which marks off the natures of things, as though the void were some sort of separator and determiner of things next to each other. And this applies first of all to the numbers, for it is the void that marks off their natures.

The things on the basis of which some say that it is and others that it is not are pretty much of this kind and this many.

Chapter 7

30 To move toward which way it is, one must take up what the word means. And the void seems to be a place in which nothing is. The reason for this is that people think what *is* is body, and every body is in a place, and a place in which there is no body is void, *214a* so that if somewhere there is no body, there is nothing there. They suppose in turn that all body is tangible, and the tangible is what has heaviness or lightness. It follows then by reckoning these together that this is the void: that in which there is nothing heavy or light. But while this follows, as we said before, deductively, it would be absurd if a point were said to be a void, for it must be a place in which there is extension for a tangible body. So it is evident that the void is meant in one way as what is not full of body sensible to touch, and *10* sensible to touch is what has heaviness or lightness. (And because of this one might produce an impasse: What would they say if the extension contained color or sound? Would it be a void or not? It is clear that if it could receive a tangible body it would be a void, but if not, not.) But in another way, a void is that in which there is no *this*, nor any bodily being. For this reason, some say that the void is the material *of* body (these also say that place is this same thing), but they do not speak well, for the material is not separable from the thing, but they inquire after the void as separable.

But since we have distinguished what place is, and the void, if there is one, must be a place deprived of body, and in what way a place is and in what way it is not have been *20* said, it is clear in this way that there is no void, either separated or inseparable. For the void is meant to be not body but an extension of body. And hence the void is thought to be something, since place is, and for the same reasons. For motion with respect to place comes in both for those who say that place is something apart from the bodies that happen to be in it, and for those who say that the void is. They think the void is a cause of motion as that in which something is moved, and this would be the sort of thing some people say place is. But there is no necessity at all that if there is motion, there be a void. In general, a void is in no way necessary for every motion, on account of something which escaped the notice of Melissus; for what is full admits of alteration. But neither is a void necessary for change of *30* place, for it is possible for things simultaneously to make way for one another, there being no separate extension apart from the bodies moved. And this is clear also in the whirling motion within continuous things, just as in those of liquids. And it is possible for something *214b* to be compressed not into the void but by squeezing out the pits within it (as water is compressed because of the air within), or to be increased not only by the entrance of something but also by alteration, for example if air were to come into being out of water. And in general, both the argument about increase and the one about pouring water into ashes step on their own feet. For either not any part whatever increases, or not by means of body, or it is possible for two bodies to be in the same place (then they would be claiming to resolve a common impasse, but not proving that there is a void), or the whole body must *10* be a void if it increases in every part and increases by means of void. And the same argument applies also to the ashes. That, then, the things from which they prove that there is a void are easy to undo is clear.

Chapter 8

Let us say again that there is no separate void, as some say there is. For if there is some change of place of each of the simple bodies by nature, as with fire up and with earth down and toward the center, it is clear that the void could not be responsible for the change of place. For what, then, would the void be responsible? For it seems to be a cause of motion with respect to place, but of this it is not a cause. Further, if the void is such a thing as a place deprived of body, then when there is a void, where will a body placed into it be *20* carried? Surely not into all of it. But the same argument goes toward those who think that place is something separate into which a thing is carried: in what way will the thing placed in it be carried or stand still? And the same argument about up and down will also fit the void, appropriately, since those who say there is a void make it a place. And how, to begin with, will something *be* in a place or in the void? For it does not follow, whenever some body as a whole is set down, that it is in a separate and persisting place, for the part, if it is not set down separately, will not be in a place but in the whole. And if there is no place for it, neither will there be a void.

30 And to those who say there must be a void if there is to be motion, it turns out rather to be the other way around when one examines it, that it is not possible for even one single thing to be moved if there is a void. For just as people say that the earth is at rest because of its being evenly balanced, so also is it necessary to be at rest in the void: for there is nowhere that *215a* a thing will be moved more or less than anywhere else, since insofar as it is void, it has no differences. Besides, every motion is either forced or by nature. And necessarily, if there is a forced kind, there is also the natural kind (for what is forced is contrary to nature, and what is contrary to nature is secondary to what is by nature). So if there is not a motion in accordance with nature for each of the natural bodies, neither will there be any at all of the other motions. But how will there be a motion by nature if there is no differentiation throughout the void or the infinite? But insofar as it is infinite there will be no up, down, *10* or center, and insofar as it is void, up is no different from down (for just as within nothing there is no difference, so also within the void, for the void seems to be some sort of non-being and deprivation). But change of place by nature is differentiated, so there is difference by nature. Then either there is no change of place by nature anywhere for anything, or if there is, there is no void.

Further, as things are, things that are thrown are moved when they are not touching what pushed them, either because of circular replacement,° as some say, or because the air pushes the thing that was pushed with a motion faster than that by which it is carried to its proper place. But in a void, none of these things is possible, nor will anything be moved *20* except by being carried by something that holds it up. Moreover, no one could say why a thing moved would stop anywhere, for why here rather than there? So it will either stand still or it must be carried into infinity, unless something stronger gets in its way. Again, it seems now that a thing is carried into the void because of its yielding, but in the void such a thing happens equally in every direction, so the thing will have been carried in every direction.

And yet further, what is being said is clear from the following things. For we see that the same weight and body is carried faster through two causes, either by a difference in the through-which, as through water rather than earth or through air rather than water, or by a difference in what is carried, if the other things are equal, because of an *30* excess of heaviness or lightness. That through which it is carried is a cause which impedes it most when *it* is carried in the opposite direction, but secondly also when it is still, and *215b* more so that which is not easy to divide, and this is what is thicker. So A will have been carried through B in the time C, but through D, which is thinner, in the time E, if the distance through B is equal to that through D, the times being in the same ratio as the impeding bodies. For let B be water and D air; by as much as air is thinner and less bodily than water, by this much will A have been carried faster through D than through B. Let the speed have to the speed the same ratio that stands between air and water. So if the first *10* is doubly thin, it will go through B in a time double that through D, and time C will be the double of E. And always, by so much as that through which it is carried is less bodily, less resistant, and more easily divided, the faster will it have been carried. But the void has no ratio by which it comes short of body, just as also the nothing has none to a number. For if four things exceed three by one, and exceed two by more, and one by still more than two, to the nothing they no longer have a ratio by which they exceed. For the thing that exceeds must be divisible into the excess and that which is exceeded, so the four will be as many as it exceeds by together with nothing. For which reason a line also does not exceed a point, *20* unless it is composed of points. Similarly, the void can have no ratio to the full, so neither can the motions through them have a ratio, but if through the thinnest thing such a thing is carried in so much time, through the void it outstrips every ratio.

For let F be a void, equal in magnitude to B and D. Now if A goes through it, and will have been moved in some time G, less than E, the void will have this ratio to the full. But in a time just as much as G, in D, A will go through a distance H. But it will also go *30* through some thing F which is different from air in thinness in the same ratio which the time E has to G. For if the body F is thinner than D by so much as E exceeds G, A will go *216a* through F with a speed determined by the inverse ratio of the times, if it is carried there. So if there is no body in F, it will go still faster. But it went through it in time G. Therefore, in an equal time it goes through what is full and what is void, which is impossible. Accordingly, it is clear that if there is a time in which a thing will be carried through any amount whatever of void, this impossible thing will follow: an equal time will have been taken to traverse something which is full and something which is void. For there will be some body in the same ratio to the other as the time to the time. And to state the main point, the cause of this result is clear, that of every motion to any other motion there is a *10* ratio (for they are in time, and of every time there is a ratio to any other time, if both are finite), but of a void to the full there is none.

These things follow insofar as those things through which bodies are carried differ, but the following things from an excess in what is carried. For we see that things which have a greater preponderance of heaviness or lightness, other things being equal, are carried faster through an equal interval, and in accordance with the ratio the magnitudes have to one another. And so also through the void. But this is impossible, for through what cause will it have been carried faster? In what is full it is by necessity, for the greater divides it *20* faster by means of its strength. For what is carried or let go divides either by means of its shape or by means of the preponderance it has. Therefore in the void all things will be of equal speed, which is impossible.

It is clear, then, from what has been said, that, if there is a void, there follows the opposite of that on account of which those who say there is a void make it up. Some think that if there is to be motion with respect to place, there is a void set off by itself, but this is the same thing as saying that place is something separated, and that *this* is impossible was said before.

But even in its own right, the so-called void would seem to those who examine it to be a truly empty idea. For just as, if someone places a cube in water, it will displace an amount *30* of water that is as much as the cube, so also is it in air, though it is not evident to the senses. And always in every body capable of displacement, in the direction in which it is its nature to be displaced, it is necessary, if it is not compressed, that it be displaced, either always downward, if, like that of earth, its motion is downward, or upward, if it is fire, or in both directions, whatever sort of thing might be placed in it. But surely in the void this is impossible (since *216b* it has no body), but there would be dissolved through the cube an interval equal to that which was in the void before, just as if the water had not been displaced by the wooden cube, nor the air, but went all through it everywhere. But surely the cube has as much magnitude as it occupies of the void; which magnitude, if it is also hot or cold, or heavy or light, is nonetheless different in what it is from all its attributes, even if inseparable from them. I mean the bulk of the wooden cube. So even if it were separated from all the other things, and were neither heavy nor light, it would occupy the equal void and be in the same *10* place as that part of the place or of the void equal to itself. In what way, then, will the body of the cube differ from the equal void or place? And if there are two such things, why would there not be any number whatever in the same place? So this is one absurd and impossible thing. But still it is clear this cube will also have a displacement, which all other bodies have. So if this does not differ from its place, why is it necessary to make a place for bodies apart from the bulk of each one, if the bulk were without attributes? For it contributes nothing if there is another such equal interval in it. {Yet such a thing as void ought to be evident among moving things, but it is nowhere in the world now. For the air is something, though it does not seem to be—nor would water if fish were iron; for the discrimination of the tangible is by the sense of touch.} *20* That, then, there is no separated void, is clear from these things.

Chapter 9

There are some who think it is clear there is a void on account of the rare and the dense. For if there were not rare and dense, neither would it be possible for anything to contract or be compressed; and if this were not possible, either there would be no motion altogether, or the whole would swell up, as Xuthus says it does, or air and water would always change into equal amounts (I mean, say, if air has come into being from a cup of water, at the same time from an equal amount of air, just so much water would have come *30* into being), or there must necessarily be a void. For to be compressed or to be expanded would not be possible otherwise. If, then, by the rare they mean that which has many separated voids, it is clear that if there cannot be a separate void, nor likewise a place having its own extension, neither is there anything rare in that way. But if, though it is not separate, there is nevertheless some sort of void present in the rare, this is less impossible, but, first *217a* of all, it follows that the void would not be responsible for every motion, but only the upward (for the rare is light, on account of which people say fire is rarefied), and also, the void would be a cause of motion not as that in which it happens, but just as wineskins, by being carried upward themselves, carry what is connected to them, so would the void carry things up. Yet how can a void have a change of place, or have a place? For a void then comes to have a void, into which it is carried. Further, how will they give an account for the heavy of its being carried downward? And it is clear that if a thing will be carried upward according to how much the more rare and more void it is, if it were completely void it would be carried fastest. But perhaps this cannot even be moved, but rather the same *10* argument that showed that in the void all things are immovable also shows in the same way that the void is immovable, for the speeds cannot be made sense of.

But since we say there is no void, the other things are truly at an impasse: that either there would be no motion if there were no becoming-dense or becoming-rare, or the heaven would swell up, or there would always be an equal amount of water from air and air from water. (For it is clear that a greater amount of air comes into being out of water; so it is necessary, if there is no compression, either that what is next to it be forced outward to make the extremity swell out, or that somewhere else an equal amount of water be transformed out of air, in order that the entire bulk of the whole be equal, or that nothing be moved. For these alternatives will always follow when anything is displaced, unless it *20* comes around in a circle; but change of place is not always in a circle, but also in a straight line.) So on account of these things, some would say there is some sort of void, but we say that, from among the underlying things, there is one material of contraries, of hot and cold and of the other natural opposites, and from what *is* potentially comes what is at work, while the material is not separate, though the being of it is different, and is one in number, as it might happen to skin to be the material of both hot and cold.

But there is also the same material of both a large and a small body. For clearly, whenever air comes into being from water, the same material became it without receiving *30* anything else in addition, but what it was potentially it became actively, and in turn water comes from air in the same way, turning at one time into largeness out of smallness, at another into smallness out of largeness. And in like manner, even if the air, being much, becomes less in bulk, or from less greater, the material which was potentially both becomes both. For just as the same thing turns from cold to hot and from hot to cold because it was *217b* so potentially, so also it turns from hot to more hot, though nothing in the material becomes hot which was not hot when it was less hot; just as neither is it the case, if the circumference and curvature of a larger circle were to become those of a smaller circle, whether the same one or another, that curvature would have become present in anything which had not been curved but straight (for it is not by leaving gaps that there is a less and a greater). Nor is it possible to take any magnitude of a flame in which there is not present both heat and brightness. So too is the earlier heat all through the later heat. So also the largeness and smallness of sensible bulk are expanded not because the material receives anything in *10* addition, but because the material is potentially both; so the same thing is dense and rare, and there is one material of them.

But the dense is heavy and the rare light. Yet just as the circumference of a circle when drawn together into a smaller one does not receive some other concave thing, but what was present is drawn together, and every bit of fire whatever that one might take will be hot, so also in general is there a drawing together and pulling apart of the same material. And there are two aspects each of the dense and the rare, for both the heavy and the hard seem to be dense, and their opposites rare, both the light and the soft. But the *20* heavy and the hard are not in unison in lead and iron.

From what has been said it is clear that there is no distinct void, neither simply, nor in what is rarefied, nor potentially, unless one wants to call the cause of being moved in general "void". But in that way, the material of the heavy and light, as such, would be the void. For the dense and the rare, by virtue of *that* opposition, are productive of being affected or unaffected, and of qualitative change rather than of change of place. And as for the void, in what way there is one and in what way there is not, let it have been marked out in this way.

Commentary on Book IV, Chapters 6–9

The clepsydra ("water thief") was a hollow container with holes in the bottom and a tube coming out of the top. It was used as a water clock, but also to draw wine or other liquids from vats too large to pour from. Submerge the clepsydra, let it fill, stop up the tube with a finger or thumb, lift it out, and the liquid will not run out. This was one of many ancient experiments interpreted as showing that there is no void. It is cited in the fourteenth century by Marsilius of Inghen (*Questions on the Eight Books of the Physics*) as proof that "nature abhors a vacuum." Aristotle says that such arguments show no such thing, and do not even approach the doorway in front of the question about the void, but only demonstrate the strength the air has to overpower even the weights of heavy bodies. This is the explanation given for these effects in the seventeenth century by Pascal, who thus refuted a long tradition but was only agreeing with Aristotle.

Aristotle considers the arguments made in favor of the void a better starting place for the inquiry. People say that a body could never move into what is full, but only into a void. If a place already full could accept any new body, the whole world could be packed into it. Also, since many bodies can be compressed, they must have voids in them to start with; and a jar full of ashes will hold just as much water as the same jar will when empty. Aristotle replies that a body could move into what is full if, at the same time, another body is displaced from it; stir a liquid, and its parts will whirl around in a continual exchange of places. This sort of circular replacement occurs when a valve is opened at the bottom of a wine barrel; if the barrel is otherwise sealed, nothing will come out, but if there is an opening for air to enter, the wine will flow out. Similarly, a body can be compressed by squeezing out pockets of air, just as squeezing half a lemon makes the pits fly out, and the ashes in the jar might be almost all air.

The void could not cause motion, Aristotle points out, just because it is homogeneous and undifferentiated. What sufficient reason would there be for a body to go in any direction in particular? And when a body travels through any actual medium, it goes faster, other things being equal, the thinner and less resistant the medium is. But if it goes five miles an hour in one medium, and ten miles an hour in a medium half as resistant, how fast would it go in the void? Aristotle says that, since there is no ratio of the thickness of void to that of any real medium, the question admits of no intelligible answer. Putting the question another way, Aristotle asks what would happen if several bodies of different shapes and weights were let go in a void. Since there is nothing to be divided by a more "aerodynamic" shape, and no resistance to be overcome by a greater weight, all bodies would move at the same speed. Aristotle considers this a second self-contradiction contained in the idea of motion in a void.

We, of course, have been taught that things do all fall at the same speed in a void. This criticism of Aristotle goes back to the sixth century, when John Philoponus argued that a body would have a definite speed in a void, determined by the excess of the strength that set it moving over the strength of the resistance. Since no ratio of zero to a magnitude is required, but only an amount by which a magnitude exceeds zero, the result is intelligible, and any two bodies propelled in the void with equal force would have equal speeds, no matter how they differed in other respects. This conception does refute Aristotle's objections to motion in a void, but at a high price. The argument of Philoponus contains elements of Newton's first and second laws of motion: that in the void the body would go on moving when no cause was at work on it, and that a violent push imposed on a body would be internalized into it, no matter how opposed that violent motion might be to the natural motion of that kind of body. The mathematical arguments in IV, 8, do not stand in isolation, but within the context of a prior conception of nature and of being. If there is motion in a void, there is inertia; if being is being-at-work, there is no motion in a void.

A body in a surrounding medium always displaces a part of that medium equal to itself in bulk. In a void that same bulk would be occupied, but not displaced. But this is just a way of saying that a bulk of a certain amount is one of the attributes that belong to the body. To say that it is in an equal volume of void is a confusion by which that attribute is posited as having a separate existence. It is the same confusion that produces the concept of space.

But if everything is continuous and full, how does a body expand or contract? Aristotle says this is no different from asking how a body becomes hotter or colder, or a circular arc becomes more or less curved. There is, he says, always one and the same material for a pair of contraries. The arc can be pulled into a tighter circle; it doesn't need to have new pieces of concavity added to it. Similarly, a body can contract or expand while remaining continuous throughout, without subtracting or adding pieces of void. In the geometrical imagination, we can form no picture corresponding to this increase or decrease in density, but how did the imagination become the ultimate judge of what does and what does not happen in the world? The last chapter of Book III illustrates the absurdities that lie along that path. We are more comfortable picturing particles moving closer together or farther apart in a void, but this is a false comfort. How do the particles themselves possess their densities? Ultimately there must be something continuous and free of void.

With a prior conception of being as being-at-work, and with a critical attitude that does not accept the seductive testimony of imagination at face value, Aristotle has a solid argument against the possibility of void. That argument completes the sequence of inquiries into the infinite, place, and the void. One condition of motion, as Aristotle understands motion, is a finite place, free of void.

Book IV, Chapters 10–14

Time

Chapter 10

217b, 29 To attack the subject of time is connected with the things that have been said. And first, it is a good idea to come to an impasse about it by way of the popular arguments about whether it is something that is or something that is not, and then about what its nature is. That time either has no being at all, or *is* only scarcely and faintly, one might suspect *218a* from this: part of it has happened and is not, while the other part is going to be but is not yet, and it is out of these that the infinite, or any given, time is composed. But it would seem impossible for a thing composed of non-beings to have any share in being.

Further, if any composite thing is to be, it is necessary that while it is, all or some of its parts must be; but though time is composite, part of it has happened and part is going to be, while none of it is. The *now* is no part of it. For the part measures° the whole, and *10* the whole must be composed of the parts, but time does not seem to be composed of nows. Also, it is not easy to see whether the now which divides the past and the future remains one and the same, or is always another and another. For if it is always other and other, while no parts of time which are other and other can *be* at the same time (unless one of them surrounds or is surrounded, as the lesser time is by the greater), and if the previous now is not, but is necessarily then annihilated, then the nows will not be at the same time as one another, but the previous one must always be annihilated. And in itself it is not possible for it to be annihilated, since then it is, but the previous now is not capable of being annihilated in another now. For take it as impossible that the nows be touching each other, just as it is impossible that a point can touch a point. Then if it is annihilated not in the succeeding now [since there is no next succeeding one], *20* but in another, then it would *be* at the same time as the infinitely many nows between them: but this is impossible. But surely neither is the always-remaining-itself possible. For no determinate thing can terminate in one limit, whether it is continuous in one or more than one direction. But the now is a limit, and time can be taken as limited. Further, if being coincident in time, and neither before nor after, means to be in one and the same now, and if the things before and the things after are in this now, then the things that happened ten thousand years ago would be at the same time as the things happening today, and neither before nor after would be anything distinct from the other.

30 Let these be a sufficient number of impasses about the things belonging to the subject. But what time is, and what its nature is, is just as unclear from the things handed *218b* down to us, as are also the matters we have just taken up one by one. For some say that time is the motion of the whole, others that it is the celestial sphere itself. But surely a part of a circuit is a certain time, though it is not a circuit; for the thing taken is part of a circuit, but not a circuit. Moreover, if there were more heavens than one, the motion of any of them whatever would likewise be time, so that there would be many times at the same time. And to those who say that the sphere of the whole is time, this seems so because everything is in time, as well as in the sphere of the whole. But this is too simple-minded a proposition to need examination of the impossibilities that follow from it.

But since time seems most of all to be a certain motion and change, this is something *10* one must examine. Now the change or motion of each thing is only in the changing thing itself, or in the place where it happens to be moving or changing, but time is present in the same way everywhere and to all things. And further, change is faster or slower, but time is not. For the slow and the fast are defined by means of time, the fast as what is moved much in a little time, the slow as what is moved little in much time; but time is not defined by means of time, neither by being a certain amount of it nor a certain kind. That, then, time is not motion is clear, and it makes no difference to us in the present inquiry to speak of motion or of change.

Chapter 11

20 But neither is there time without change. For whenever we ourselves are not changing at all in our thinking, or are unaware of our changing, it does not seem to us that time has happened, as when those people wake up who are said in stories in Sardinia to sleep among the heroes. For they join together the earlier now and the later now and make them one, taking out the in-between on account of being unconscious. Then just as if the now were not other, but were one and the same, there would be no time, so also when we are unaware of its being other, there does not seem to be an in-between time. But if not supposing time *30* to be happens to us sometimes, whenever we do not mark off any change but the soul seems to remain in a state that is one and indivisible, while whenever we do perceive and mark it off, we then say that time has happened, it is clear that without motion and change, time is not. *219a* That, then, time neither is motion nor is without motion, is clear. And since we are seeking what time is from its sources, we must take up what motion is. For we perceive time together with motion, and even if it were dark and we experienced nothing through the body, but a certain motion were present in the soul, immediately, a certain time would seem to happen along with it. Indeed, whenever any time seems to happen, some motion seems also to happen. Therefore surely what time *is* either is motion or is *of* motion. Then since it is *10* not motion, it is necessary that it be something of a motion.

Now since a moving thing is moved from something to something, and every magnitude is continuous, the motion follows the magnitude. For through the magnitude's being continuous, the motion too is continuous, and through the motion the time. For as much as the motion is, so much also does the time always seem to become. And before and after belong first of all to place, and thereby to position. But since there is a before and an after in the magnitude, it is necessary that there also be a before and an after in motion, *20* analogous to those in the magnitude. But also there is a before and an after in time, since the one of them always follows the other. And whenever there is a motion, there is a before and an after in it. But surely being before-and-after is something else and not motion. But we recognize time whenever we mark off a motion, marking it off by means of a before and an after. And that is when we say that time has happened, whenever we take cognizance of the before and after in a motion. And we mark them by taking them to be other and other with something else between them. For whenever we think the extremes differ from the middle, and the soul says there are two nows, one before and one after, then also we say this to be time. *30* For time seems to be bounded by the now; and let this be laid down. So whenever we preceive the now as one, and neither as before and after in a motion, nor the same but belonging to something before and something after, no time seems to have happened because *219b* no motion has. But whenever there is a before and an after, then we say there is time, for this is time: a number of motion fitting along the before-and-after.

Therefore time is not motion except insofar as the motion has a number. Here is a sign of this. We judge the greater and less by number, and we judge a motion to be greater or less by time; time therefore is some sort of number. But since the meaning of number is twofold (for we call that which is counted or countable a number, but also that by which we count), time is the counted and not the number by which we count. The two senses *10* of number are different. And just as motion is always other and other, so also is time. (All times that coincide are the same. For the now is the same at any one time, though the being of it differs. But the now divides the time into a before and an after.) But the now is present in one way as the same, but in another as not the same. Insofar as it is in something other and other, it is different (and this is what it means to be the now), but whatever, being at any one time, is the now, is the same. For as was said, the motion follows a magnitude, and as we say, the time follows the motion. And it is likewise with the thing carried along in relation to the point, by which we recognize the motion and the before and after in it. *20* This at any one time is the same (for it is either a point or a stone or some other such thing), but it differs in meaning, just as the sophists take as different Coriscus in the Lyceum and Coriscus in the agora. And this, by its being elsewhere and elsewhere, is different; but the thing carried along corresponds to the now, just as the motion to the time. (For it is by the thing carried along that we recognize the before and after in a motion, and insofar as the before and after are countable, that by which we recognize them is the now.) So also among these, that which, being the now at any time, is in one way the same (for it is what is both before and after in the motion), is in another way different (for insofar as the before and after are countable, the now divides them). And it is this that is most knowable; for the *30* motion is known through the thing moved, and the change of place is known through the thing carried along. For the thing carried along is a *this,* but the motion is not. So the now exists in one way as always the same, but in another way as not the same, for so also does the thing carried along.

And it is clear both that, if there were no time, there would be no now, and, if there *220a* were no now, there would be no time. For just as the thing carried along and the change of place are at the same time, so also are the number of the thing carried along and that of the change of place. For the number of the change of place is time, and the now is manifest as the thing carried along, like a unit of number. But in addition [unlike number], time is continuous by means of the now, and is divided by the now. This also follows the change of place and the thing carried along. For also the motion and the change of place are one in the thing carried along, because it is one (not just by *being* at one time, since it might make stops, but in meaning). And the now marks off the motion into a before and an after; *10* and this it does in a manner corresponding to that of the point. For the point also both holds together the length and divides it, since it is a beginning of one part and an end of another. But whenever one takes it in this way, using what is one as two, it is necessary to make a stop, if the same point is to be a beginning and an end. But the now is always other through the moving of the thing carried along. So time is a number not as of the same point, which is a beginning and an end, but rather as the extremities of a line; and neither is time a number as parts, both because of what has been said (for one might use the intermediate point as two, so that time would happen to stand still), and further because it is clear that the now is no part of time, nor is the division part of the motion, just as neither is *20* the point any part of the line, but it is two lines that are parts of one line. Then insofar as the now is a limit, it is not time but an attribute of time; but insofar as it numbers, it is a number. For the limits belong only to that of which they are the limits, but the ten which is the number of these horses belongs also elsewhere. That, then, time is a number of motion fitting along the before and after, and that it is continuous (for it is of something continuous), is clear.

Chapter 12

The smallest number, simply, is two; but it is possible with some particular kind of counted thing that in one sense there is a smallest number, in another not. For example, of a line, the smallest number in multitude is two, or one, but in magnitude there is no smallest. For *30* every line is always divisible, and so likewise is time. For the smallest amount of it in number is one or two, but in magnitude there is no smallest.

220b And it is clear that time is not said to be fast or slow, but much or little, or long or short. Insofar as it is continuous, it is long or short, but insofar as it is a number it is great or small. Fast or slow it is not, for neither is the number by which we count anything fast or slow.

And time is the same everywhere, at the same time. But that before and that after are not the same. This is because, though the present change is one, the change that has happened and the one that is about to be are different. But time is a number of change, *10* not that by which we count but that which is counted, so with this also it happens that the before and after are always different; for the nows are different. And the number is one and the same of a hundred horses and of a hundred human beings, but those things of which it is a number are different, the horses and the human beings. Also, just as it is possible for a motion to be one and the same again and again, so also a time can be, such as a year or spring or fall.

And not only do we measure a motion by a time, but also a time by a motion, since they mark off one another's boundaries. For the time marks off the motion as a number of it, but the motion marks the time. And we call a time much or little, measuring it by the *20* motion, exactly what we do when we determine a number by the thing numbered, as the number of horses by means of the one horse. For we recognize the multitude of the horses by the number, and conversely, we recognize the number itself of the horses by the one horse. And similarly with time and motion: by the time we measure the motion, and by the motion the time. And this happens reasonably, for the motion follows the magnitude, and the time follows the motion, in being this much and continuous and divisible. For since the magnitude has these attributes, the motion receives them, and through the motion the time receives them. And we measure the magnitude by the motion and the motion by the magnitude. When *30* there is a lot of walking we speak of a long road, and we call the walking a lot when the road is long; also the time is long when the motion is much, and the motion much when the time is long. *221a* And since time is a measure of motion and of being moved, and measures the motion by marking off a certain motion which measures out the whole into pieces (just as also the cubit measures a length by marking off a certain magnitude which measures out the whole); and since for the motion to be in time means that both it and the being of it are measured by time (for it measures at the same time both the motion and the being of the motion, and this is its being-in-time: the being-measured of the *being* of it); it is clear that for other things too this is the being-in-time: the being-measured of the being of them by time. *10* For to be in time is one of two things: to be when time is, or something corresponding to what we say of some things, that they are in number. This means that something is a part of or attribute of or generally something belonging to number, or else that there is a number of it. And since time is a number, the now and the before and such things are in time in the same way that one and the odd and even are in number (for each of the latter is something belonging to number and each of the former is something belonging to time). But things are in time as in number, and if this is so, they are surrounded by time just as things in *20* place are by place. And it is clear that to be in time is not to be when time is, any more than to be in motion or in place is to be when motion or place is. For if that is to be in something, all things will be in anything, and the heavens in a millet seed. For at the time when the millet seed is, the heavens are also. But this is incidental, while the other follows necessarily, both for the thing in time that there be a certain time when it is, and for the thing in motion that there be then a motion.

Since being in time is like being in number, some time may be taken greater than anything which is in time; on account of which all things *in* time must be surrounded by time, *30* as are any other things in something, as things in place are surrounded by place. And a thing is acted upon by time, as we are accustomed to say that time wastes things away, or that everything ages by the action of time or forgets because of time, but not that anything *221b* has learned or become young or beautiful because of it. For time in itself is a cause rather of destruction; for it is a number of motion, and motion displaces what is laid down. So it is clear that things that always are, insofar as they always are, are not in time; for they are not surrounded by time, nor is the being of them measured by time. A sign of this is that they are not affected by time in any way, just as if they were not in it.

And since time is a measure of motion, it will also be a measure of rest; for every act *10* of resting is in time. For though a thing in motion must be moved, this is not so for a thing in time; for time is not motion, but a number of motion, and it is possible for a resting thing to be in a number of motion. For not every motionless thing is at rest, but only a thing deprived of motion which is by nature such as to be moved, as was said before. And for a thing to be in number is for there to be some number of the thing, and for the being of it to be measured by the number in which it is, that is by time, if it is in time. And time will measure the moving thing and the resting thing, the one insofar as it is moving, the other insofar as it is resting; for it will measure the motion and the rest of them as a certain how-much. So the moving thing will not simply be measured by time insofar as it is so much, but *20* insofar as the motion of it is so much. So as many things as neither move nor rest are not in time; for to be in time is to be measured by time, and time is a measure of motion and rest.

It is clear, then, too, that not every non-being will be in time, for example those that cannot be otherwise than non-beings, such as a diagonal of a square which is commensurable with its side.° For generally, if time in itself is a measure of motion, but measures other things only incidentally, it is clear that for all those things of which it measures the being, that being will be in rest and motion. As many things as pass away and *30* come into being, and generally are at one time but at another time are not, must be in time (for there is some greater time which exceeds both the being of them and the time measuring that being). And of the non-beings that time does surround, some were, *222a* as Homer once was, and others will be, such as anything that is about to be, and time encloses them on at least one side; and if time encloses them on both sides, they both were and will be. As many things as time in no way surrounds neither were nor are nor will be. Among the non-beings, such will be as many as have opposites which always are, as the being-incommensurable of the diagonal always is, and these will not be in time. So the commensurability of it will not be in time, since it is always a non-being because its opposite always *is.* As many things as do not have opposites which always are, are capable both of being and not being, and it is of them that there is coming-to-be and passing away.

Chapter 13

10 The now is a connection of time, as was said; for it connects the past time and the future. And it is a boundary of time; for it is the beginning of one part and the end of another. But this is not as clear as with the point, which stands still. And the now divides, potentially; and insofar as it is a division, the now is always different, but insofar as it binds together, it is always the same, as with mathematical lines. (For a point is not always the same in our understanding; for as dividing, it is other and other, but insofar as it is one, it is the same in every respect.) Thus also, the now is in one way a division of time, potentially, and in another a common boundary and union of its parts. And the dividing and *20* the uniting are the same act of the same thing, but the being of them is not the same.

In one sense, then, the now is meant in this way, but it is meant in another way whenever the time is close to this one. "He will come now," because he will come today; "he has come now," because he came today. But the things in the *Iliad* did not happen just now, nor did the great flood. Surely the time back to them is continuous, but they are not nearby.

"Sometime" is defined in relation to the former kind of now, as in "Troy was captured sometime," or "sometime there will be a flood." It is necessary that these extend right up to the now in question; and therefore there will be a certain amount of time from this one to that one, or there was, if it is a past time. And if there is no time which is not "sometime," then every time will be of a definite extent.

30 Will time then come to an end? Or will it not, if, that is to say, there is always motion? And then is it different, or the same many times? It is clear that as the motion, so also the time; for if it sometimes becomes one and the same, the time also will be one *222b* and the same, but if not, it will not. And since the now is an end and a beginning of time, but not of the same time, but an end of the past and a beginning of the future, it would be like the circle, which is in a certain respect in the same place both convex and concave; so also is time always at a beginning and an end. And because of this it seems always different, for it is not of the same thing that the now is a beginning and an end. If it were, it would be opposite things at the same time and in the same respect. And it will not come to an end, for it is always at a beginning.

"Right away" is the part of future time which is near the indivisible present now. ("When are you walking?" "Right away," because the time in which it will be is near.) And *10* "just now" is that part of the past time which is not far from the now. ("When are you walking?" "I walked just now.") But to say that Troy was just now conquered is not the way we speak, because it is too far from the now. Also, "lately" is a part of the past near the present now: "When have you come?" "Lately," if the time were near the now that is at hand. But "long ago" is the far away.

The sudden is what is displaced in a time imperceptible on account of smallness, but every change is by nature a displacement. In time everything comes into being and passes away, on account of which some used to call it the wisest thing, but the Pythagorean Paron, speaking more rightly, called it the stupidest thing, since in it things are forgotten. So it is *20* clear that in itself it will be a cause rather of destruction than of generation, exactly as was said before (for change is in itself a displacer), but incidentally of generation and of being. A sufficient sign of this is that nothing comes into being without itself somehow moving and acting, but something can be destroyed and in no way moved. And this is what we are most of all accustomed to mean by the destroying action of time. But surely not even this does time *do*, but even this change happens in time incidentally.

That, then, time is, and what it is, and in how many ways the now is spoken of, and what sometime, lately, right away, just now, long ago, and suddenly mean, have been said.

Chapter 14

30 These things having been determined thus for us, it is clear that every change and everything moved are in time. For faster and slower are applicable to every change (for *223a* in every one of them it is obviously so). By being moved faster I mean changing sooner into whatever is proposed, along the same interval, and moved with a uniform motion (as in change of place, if both things are moved along a circumference or both along a straight path, and similarly with other things.) But surely the "sooner" is in time, for we speak of before and after in relation to the separation from

the now, while the now is a boundary of the past and the future. So since the nows are in time, the before and after will also be in time; for in that in which the now is, the separation from the now is also. (The before is spoken of in opposite ways in relation to the past time and the future. For in the past we 10 call what is farther from the now before and what is nearer after, but in the future, we call the nearer before and the farther after.) So since the before is in time, and the before follows along with every motion, it is clear that every change and every motion are in time.

Worthy of examination, too, is how time stands in relation to the soul, and why time seems to be in everything, the things in the earth and in the sea and in the heavens. Is it because it is an attribute or state of motion, since it is a number of it, while all these are *20* movable things (for all are in place), while time and motion go together both in the course of potentiality and in being-at-work?

And one might be at an impasse whether, if the soul were not, time would be or not. For if it is impossible for there to be a counter, it is also impossible for there to be anything counted; so it is clear that neither can there be a number. For a number is either what has been counted or what is capable of being counted. But if nothing else is of such a nature as to count but the soul and the intelligence in the soul, then it is impossible that time be if soul is not; but there would still be that which time depends on the being of, such as motion, if it is possible for *it* to be without soul. But the before and after are in a motion, and insofar as they are counted, they are time.

30 And one might be at an impasse also about what sort of motion time is a number of. Or is it of any kind whatever? For in time a thing comes into being and passes away and grows and alters and changes place. Insofar as anything is a motion, then, time is for this *223b* reason a number of each motion. On account of which, it is simply a number of continuous motion, not of a certain kind. But it is possible for something else also to be moved now, and of each of the two motions there would be a number. Is there then another time, and would there be at the same time two equal times? Or not? For a time that is equal and coincident is one and the same time. And even those not coinciding are the same in kind. For if there were dogs and horses, and seven of each, it would be the same number. So also of motions with coincident limits there is the same time, though the one perhaps is fast and the other not, or the one a change of place and the other an alteration. Surely the time is *10* the same, if it is also equal and coincident, of both the alteration and change of place. And for this reason, while the motions are different and separate, the time is altogether the same, because, of what is equal and coincident, the number is altogether one and the same.

Now since there is change of place, and included in this is motion in a circle, and each thing is numbered by some one thing of the same kind, units by a unit, horses by a horse, and in this way also time is marked off by a certain time, and just as we said, the time is measured by the motion and the motion by the time (this is because, by a motion bounded in time, the how-much both of the motion and of the time is measured); if then the first measure is a measure of all things of the same kind, uniform circular motion is a measure *20* most of all, because the number of it is most recognizable. Now neither alteration, nor growth, nor coming into being is uniform, but change of place can be. And this is why time seems to be the motion of the sphere, because by this the other motions are measured, and time by this motion. And from this comes the thing customarily said; for people say that human affairs are a circle, as well as the affairs of other things that have a nature and a coming-to-be and a passing-away. And this is because all these things are judged of by time, and come upon an end and a beginning just as if along some circuit. For also time itself *30* seems to be a sort of circle; and this in turn seems so because it is a measure of such a change of place, and is itself measured by such a one. So to say that the affairs of the things that undergo becoming are a circle is to say that time is a sort of circle; and this is because *224a* it is measured by circular motion. For apart from the measure, nothing else shows itself alongside the thing measured, but the whole is a number of measures.

It is also rightly said that there is the same number of livestock and of dogs if the two are equal, but not the same counted number ten nor the same ten things, just as neither are the equilateral and the scalene the same triangle, though surely they are the same figure, because both are triangles. For things are said to be the same in any respect in which they do not differ by any difference, but not in those respects in which they do differ, as one triangle differs from another by some difference. Therefore triangles are different from one another. *10* But they do not differ by a difference of figure, but are in one and the same division of it. For a figure of one kind is a circle, a figure of another kind a triangle, but one kind of this latter figure is equilateral, another scalene. They are the same figure, then, and this is triangle, but not the same triangle. And the number is the same (for the number of them does not differ by a difference of number), but is not the same ten; for the things of which it is said differ. Some are dogs, some horses.

About time, then, both time itself and the things related to the inquiry about it, have been discussed.

Commentary on Book IV, Chapters 10–14

Among the few and confused opinions about time that Aristotle has to start from is the belief that all things are somehow in it (218b 7). This is similar to Hesiod's image of place (208b, 29, to 209a, 3), as an empty abyss which came into being before anything else so all things would have something to be in. Both opinions correspond to Newton's idea of absolute space and time, as pre-existing homogeneous empty containers. Just as Aristotle concluded that place is not prior to bodies but dependent on them, he will conclude that time is not prior to motions but derivative from them. This is similar to the opinion of Leibniz, who considered Newtonian space and time to be fictions raised to the status of idols, confused with attributes of God himself, while "I hold space to be something purely relative, as time is" (Leibniz–Clarke Correspondence, ed. H. G. Alexander, Manchester, England: Manchester University Press, 1956, p. 25).

Aristotle's discussion of time depends heavily on certain mathematical notions that disappeared when ancient Greek mathematics was abandoned in favor of algebra, primarily through the influence of Descartes and Leibniz. Unless we recapture these notions, most of what Aristotle says about time is lost on us. His definition of time as a number of motion is meant to be a paradox, but the modern concept of (real) number conceals the difficulty in it. The mathematics of Aristotle and Euclid depends upon a distinction between magnitude (how much) and multitude (how many). Euclid, after treating ratios of magnitudes in Book V of the *Elements*, makes a new beginning to treat ratios of numbers, proving what seem to us to be the same propositions again, or at least subsets of the earlier ones. But *number* in the sense in which Euclid and Aristotle use it is always a "natural" number that counts discrete things, and can in no way be a kind of magnitude, which is always continuous. This is not just a matter of having a finicky vocabulary that insists on pedantic distinctions. Aristotle's reference to the incommensurability of the side and diagonal of a square shows the radical difference of magnitude from number. The Pythagoreans discovered and proved that no line, however small, which fits an exact number of times into the side of a square, can fit any exact number of times into its diagonal. We do not escape the difficulty by assigning to the diagonal as a "number" the square root of two, but only give it a label. The whole point is that there is no way to identify the square root of two, other than by arbitrary approximation, except as "the square root of two." The square root of sixteen has a countable, nameable meaning, but the square root of two is irrational, nameless.

Thus it is only whole numbers by which we count, but *number* in the sense in which Aristotle uses it also refers to anything which is counted. *Arithmos* primarily means not pure multitudes of intelligible units but perceptible multitudes of beings. As Aristotle said earlier (207b, 11), numbers are surnames, numbers *of* something. The dozen eggs in a carton is itself a number. But Aristotle is emphatic that time is something continuous, so how can it be a counted number? Any continuous magnitude can be turned into a number by means of measurement. To measure something, as Aristotle uses that word, means to be a proper fraction of it, to fit into it an exact number of times. The diagonal of a square is in-com-mensurable with its side because there is nothing that can measure them both. But for practical purposes we do not worry about exactness of measurement, since we are always speaking loosely anyway in assigning a number to something continuous. A length in its own right is continuous; it is measurable incidentally and arbitrarily. But time, according to Aristotle, is a continuous thing that is a number in its own right, by its very definition. It is inherently both continuous and discontinuous. And anything that is in time is not just incidentally measurable by some external procedure, but the very *being* of it is measured (221a, 8). A triangle of a certain shape, for example, may have been on the blackboard for ten minutes, or first described twenty-six centuries ago, but the being of that triangle is still atemporal. Anything can be brought in some limited way under the measurement of time, but only certain things are inherently in time, and thus measured and numbered in their being.

What are these things that are inherently temporal, and what is the measurement of them? Aristotle began his own dialectical approach to time by seeing that it is something belonging to motion, and it is as a number of motion that he defines it. And just as we measure a length by setting a short length end-to-end alongside it, so too is a motion numbered by a short motion that "measures out the whole into pieces" (221a, 1–3). Circular motions are suitable for this kind of measuring, since recurrent instances of them are easy to recognize, but any cyclical motion will do. But some third being has to do the recognizing and bring together the measuring and measured motions in one act of comparison. Since time is a number as counted, temporality is present only to a being that can count and depends on the soul (223a, 22–29). Lengths belong to bodies, and are incidentally sometimes measured by people. Time belongs not to bodies but to motions, and only insofar as the motions are brought into relation by measurement. One is not properly a number, but is what permits a numbering, so that the smallest number is two. A single motion, in isolation, lacking a number, has no time. A revery is such a disconnected motion, and we do not know if it is long or short until we leave it and look at a clock.

Time has thus only a second-order presence in the world, a presence not directly in bodies but in the comparison of the motions of bodies. Aristotle is still willing to agree with those who call time a destroyer, since coming into, and staying in, being, requires activity, being-at-work, while the mere passage of time is decay or degeneration if anything remains inert.

This section completes the analysis of the series of things implied by motion. In Book V, Aristotle's attention turns to the most general kinds of motion.

Book V

Motions as Wholes

Chapter 1

224a, 21 Either everything that changes changes incidentally, as when we say the educated one walks because that to which it is incidental to be educated walks; or it is said, because of the changing of something that belongs to it, to change simply, and such are as many things as are said to change on account of their parts (for the body is healed because the eye or the chest is, and these are parts of the whole body). But there is something that is moved neither incidentally nor on account of any other thing belonging to it, but on account of its being moved primarily. And this is what is moved in its own right, and it *30* differs in accordance with a difference in the motion, as being, say, what is altered, and within alteration it differs as what is healed or heated. And it is the same way with the mover: for one moves something incidentally, another partially, by means of moving any of the things that belong to it, and another in its own right primarily, as the doctor cures or the hand strikes. So there is something that is the primary mover, and there is something that *224b* is moved, and further an in-which, the time, and besides these a from-which and a to-which, since every motion is from something and to something. For what is primarily moved is different both from that to which and that from which it is moved, as the wood is different from both the hot and the cold. Of these, the one is that which moves, another is that to which, and the other is that from which, and it is clear that the motion is in the wood, not in its form. For the form neither moves nor is moved, nor is the place or the so-much, but there is a mover and a moved and a to-which it is moved. For the change is named by that to which it is moved rather than that from which. For this reason destruction is a change to not-being, though *10* what is destroyed also changes from being, and becoming is to being, though it is also from not-being.

What motion is was said before. And the forms and the attributes and the place, to which the moving things are moved, are unmoved, for example knowledge or heat. Someone might, however, raise an impasse, if attributes are motions, and whiteness is an attribute; for there would be a change into a motion. But perhaps it is not whiteness that is a motion, but whitening. And also in these cases there is what is incidental and what is partial and derived from something else as well as what is primary and not derived from anything else; for instance, what is whitened changes incidentally into something thought 20 about (for being thought about is incidental to the color), and it changes to color because white is a part of color (or into Europe because Athens is part of Europe), but it changes to white color in its own right. In what way, then, something is moved in its own right, and in what way incidentally, and in what way as a result of something else, and in what way by its being moved itself primarily, are clear for both what moves it and what is moved, as well as that the motion is not in the form, but in the thing that is moved or is actively movable. So let incidental change be left aside, for it is in everything and belongs to all things all the 30 time. But the non-incidental is not in everything but in contraries and what is between them, and in contradictories; and belief of this comes from examples. And a thing changes from what is in-between, since it is used as being contrary to either extreme, for the in-between in a certain way is the extremes. Hence it is also said to be in a sense contrary to them and they to it, as the middle note is sharp in relation to lowest and flat in relation to the highest, and gray is white in relation to black and black in relation to white.

225a And since every change is from something to something (as the name [*metabolē*] makes clear, for "after" [*meta*-] something else shows that there is one thing before and another after), what changes may change in four ways. For it may be from one underlying subject to another, or from a subject to what is not that subject, or from what is not a subject to that subject, or from what is not one subject to what is not another. By "subject" I mean what is declared affirmatively. So it is necessary from what has been said that there are three changes, that from one *10* subject to another, that from a subject to what is not that subject, and that from what is not a subject to that subject. For the one from what is not one subject to what is not another is not a change, on account of there being no opposition, for they are neither contraries nor contradictories. Now the change from what is not a subject to that subject, by virtue of contradiction, is coming into being, either simple, when it is simple, or some particular one, when it is of some particular thing (as the change from not white to white is a coming into being of this, but that from not-being simply, into thinghood, is simple coming into being, by which we mean that a thing comes into being simply, not that it becomes something). But the change from a subject to what is not that subject is destruction, either simply, when *20* it is from thinghood to not-being, or some particular one when it is into the opposite negation, just as was said also for coming into being.

But if what-is-not is meant in more than one way, and that which results from [faulty] combination or separation [in speech] does not admit of being moved, nor does that by way of potentiality, which is opposite to what simply is by way of activity (for the not-white or not-good nevertheless admits of being moved incidentally, since what is not white could be a human being, but what is simply not a *this* in no way admits of it), then it is impossible that what-is-not be moved. (And if this is so, it is also impossible for coming into being to be a motion, for what passes into being is what is not. For however much it becomes incidentally, it is still true to say that not-being belongs to what comes into being simply.) And it is *30* likewise impossible that what-is-not be at rest. These awkward consequences also follow if everything that is moved is in a place, since what is not is not in a place, for it would then be somewhere. And destruction is not a motion either, since the opposite of a motion is either a motion or rest, but destruction is the opposite of coming into being. And since every motion is some kind of change, and the changes mentioned are three, and of these, *225b* the ones resulting in coming into being or destruction are not motions, and these are the ones that are from a contradictory, it is necessary that the change from one subject to another be the only motion. And the subjects are either contraries or in-between (for let the deprivation also be set down as a contrary), and are declared affirmatively: naked, baregummed, or black.

If, then, the ways of attributing being are divided into thinghood, quality, place, relation, so-much, and acting or being acted upon, there are necessarily three motions, that of the of-what-kind, that of the how-much, and that with respect to place.

Chapter 2

10 There is no motion with respect to thinghood since there is not among beings a contrary to an independent thing. Nor indeed is there a motion of relation, since when one of the relative things changes, it is possible for the other, not changing at all, to be true or become untrue, so that the motion of them is incidental. Nor is there one of acting and being acted upon, or of moving and being moved, because there is not a motion of a motion, or a coming into being of coming into being, or in general a change of a change. For first of all, it might be accepted in two ways that there is a motion of a motion. It may be in the way in which there is motion of a subject. (As a human being is in motion because he is changing from white to black; is it the case that in this way also a motion heats or cools or *20* changes its place or grows or shrinks? This is impossible, since change is not any of the subjects.) Or it may be by way of the change of some other subject away from changing into another form (such as a human being from sickness to health); but neither is this possible, except incidentally, since the motion itself is a change from one form to another. It is the same situation as with coming into being and destruction, except that these are into a certain kind of opposite, but the motion is into another kind. So at the same time someone would be changing from health to sickness and from that change itself into another change. So it is clear that when he has become sick, he will be changed into whatever else the other change is into (for he could be at *30* rest), and further that is not always into what happens along, but will be from something in particular to something else, so it will be the opposite, getting well. Rather, it is through being incidental [that a change happens to a change], as there is a change from remembering to forgetting because that to which they belong changes, at one time into knowledge, at another into ignorance.

Moreover, it would proceed to infinity if there were a change of a change or a coming *226a* into being of coming into being. And an earlier one would be necessary if a later one were to be; for example, if a simple coming into being at some time came into being, then also the coming into being of it came into being, so that not yet would there be the thing that came into being simply, but a coming into being that was coming into being beforehand, and this in turn came to be at some time, so that not yet would there be the coming into being that was coming into being then. And since of infinite things there is no first one, there would not be a first becoming, and therefore no next one either, and then nothing could either come into being or be moved or change. Further, to the same thing there belongs a contrary motion (and state of rest as well), and both a coming into being and a destruction; therefore what comes into coming-into-being, whenever it comes into coming-into-being, is at that time being destroyed. For neither at the outset of its coming into being *10* nor afterward is it destroyed, since what is destroyed must *be.* Further, material must underlie both what becomes and what changes. What then would it be? Just as a body or soul is the thing that is altered, what is the thing that becomes motion or becoming? And again, what is that toward which they are moved? For motion or becoming must be of something, from something, to something. Yet at the same time, how could it be? For the process of learning could not be the coming into being of learning, so neither could becoming be the coming into being of becoming, nor anything be the coming into being of that thing. Further, if there are three kinds of motion, one of these must also be the underlying nature and that to which they are moved, for example, change of place must be *20* altered or change its place. And in general, since everything that is moved is moved in three ways, either incidentally, by means of some part, or in its own right, only incidentally would it be possible for a change to change, such as when someone being cured should run or learn; but the incidental kind we left aside long ago.

And since it is neither of thinghood nor of relation nor of acting and being acted upon, it remains that there be motion only with respect to of-what-kind, how-much, and where, for in each of these there is contrariety. Now let motion with respect to of-what-kind be alteration, for it has been joined together with this common name. But I mean by the of-what-kind not what is present in the thinghood of any thing (since then even the specific difference would be a quality), but what is of the nature of an attribute by which a thing is said to *30* be affected or unaffected. Motion with respect to how-much lacks a common name, but each-by-each it is growth and wasting away, the one into a thing's complete magnitude being growth, the one from it wasting away. Motion with respect to place is nameless both in common and in particular, but let it be called in common change of place, though only those things are properly said to be carried along which cannot come *226b* to a stop on their own when they are changing place, and which do not themselves move themselves with respect to place. And change within the same form to more or less is alteration, for it is motion either from a contrary or to a contrary, either simply or in some particular way. For what goes to the less will be said to change to the contrary, but what goes to the more will be described as changing from the contrary to the form itself. For it makes no difference whether a thing changes in some particular way or simply, except that the contraries would fall short in some way of being present; and the more or less is the greater or lesser being-present in it or being-absent-from it of the contrary.

That, then, there are only these three motions, is clear from these things. And what *10* is motionless is either that which is altogether incapable of being moved, in the way that sound is invisible; or that which is scarcely moved in a long time, or begins slowly, what is called hard to move; or that which is of such a nature as to be moved and is capable of it, but is not moved at some time when and place where and manner in which that is natural to it, the very thing which, alone among the motionless things, I call being at rest. For rest is contrary to motion, and so would be a deprivation in what admits of motion.

What, then, motion is, and what rest is, and how many changes there are, and what sort of motions, is clear from what has been said.

Chapter 3

20 After these things, let us say what the following are: coincident and separate, touching, between, next in series, next to, and continuous; and let us say to what sort of things each of these belongs by nature. Now by coincident in place, I mean whatever things are in one primary place, by separate, whatever things are in different ones, and by touching, those things of which the extremities are coincident. And the in-between is that at which a changing thing naturally arrives before it changes to what is by nature last, when it changes continuously; so the in-between involves at least three things, since it is a contrary that is the last thing in the change. And since every change involves opposites, while opposites are either contraries or contradictories, but with contradictories there is nothing in the middle, it is clear that the in-between will be among contraries. And that is moved continuously which leaves either no gap at all, or the least possible one, in the thing, not in the time (for *30* nothing precludes its having a gap left in it, or, on the other hand, the highest note's sounding immediately after the lowest) but in the thing in which it is moving. And this is evident both in changes of place and in the other changes. And what is contrary with respect to place is what is as far away as possible in a straight line, for the shortest distance is definite, and a measure is something definite.

227a And next in series is that which, being after the beginning in position or in form or in any other respect that is similarly determinate, has nothing of the same kind between it and that to which it is next in series (by which I mean, say, a line or lines if it is a line, or if it is a unit, a unit or units, or if it is a house, a house, but nothing prevents something else from being in between). For what is next in series is next in series to something, and is something that follows; for one is not next in series to two, nor the first day of the month to the second, but the latter to the former. That which, being next in series to something, *10* is touching it, is next to it. The continuous is that which is next to something, but I call them continuous only when the limits at which they are touching become one and the same, and, as the name *[suneches]* implies, hold together *[sun-echein].* And this is not possible if the extremities are two. And it is clear from this definition that the continuous is among those things out of which some one thing naturally comes into being as a result of their uniting. And in whatever way the continuous becomes one, so too will the whole be one, such as by a bolt or glue or a mortise joint, or by growing into one another.

It is clear too that the first of these relations is next in series, for what is touching is *20* necessarily next in series, but not everything that is next in series is touching. (Hence the next in series is also found among things prior in definition, for example among numbers, but contact is not.) And if things are continuous, they are necessarily touching, but if touching, not necessarily continuous; for their extremities need not be one if they are coincident, but if they are one, they are necessarily also coincident. So the last in the order of descent is what is grown into one, since the extremities must have been touching if they are to grow into one, but not everything that is touching has grown into one, though among things that have no contact, there is obviously no growth into each other either. So if a point and a unit are things of the sort that those say who claim that they are separated, it is not possible that a unit and a point be the same, since it belongs to points to be *30* [coincident], but to units to be next in series, and with distinct points there can be something in between (for every line is between points), but with distinct units, not necessarily, since nothing is between two and one.

227b What, then, coincident, separate, touching, between, next in series, next to, and continuous are, and to what sorts of things each of them belongs, have been said.

Chapter 4

Motion is said to be one in more than one way, for we mean *one* in more than one way. It is one generically according to the ways of attributing being (for a change of place is generically one with every change of place, but an alteration is generically different from a change of place), but it is specifically one whenever, being one in its genus, it is also within an indivisible kind. For example, there are differences of color—therefore turning *10* black and turning white are different in kind—but there are no more differences of whiteness, for which reason turning white is one in kind with every whitening. If there is something which is at the same time both a genus and a species, it is clear that there will be a motion that is in a way one in kind, but not one in kind simply, such as learning, if knowledge is a species of understanding but a genus of knowledges. But one might be at an impasse whether the motion is one in kind whenever the same thing changes from the same thing to the same thing, as one point changing from this place to that place again and again. But if this were so, circular motion would be the same as straight-line motion, and rolling the same as *20* walking. But are not they distinct, seeing that the motion is different if that along which it is differs in kind, while the circular path is different in kind from the straight one?

In these ways, then, motion is one in genus and species, but a motion is simply one which is one in its being and in number; and what is such is clear to those who make distinctions. For the things about which we speak in a motion are three in number, the what, the in which, and the when. And I say that there must be something that is moved, such as a human being or gold, and this must be moved in something, as in place or in an attribute, and at some time, since everything is moved in time. And of these, the being one in genus or species belongs to the thing in which it is moved, the maintaining itself to the *30* time, and the being simply one to all of these. For the in-which must be one and indivisible, as the kind, and the when, as the time, must be one and not leave gaps, and the thing moved must be one in a non-incidental way, as the white turns black or Coriscus walks (whereas the *228a* unity of Coriscus and white is incidental), nor can the thing moved be something common, since two people might at the same time be cured with the same cure, as from eye disease, but this would not be one motion, though they are one in kind. And supposing Socrates alters by an alteration that is the same in kind, but at one time and again at another, then if it were possible for something that had been destroyed to come into being again as still numerically one, this motion would also be one, but if not, the motion is the same but not one. And there is an impasse closely resembling this one: whether health is one thing, and in general *10* whether the states and attributes in bodies are one in being, for the things that have them are obviously moving and in flux. But if the health in the morning and now are one and the same, why would not this health and that be numerically one when one gains it back again after having left an interval? The argument is the same, except that it differs to this extent: if there are two states, then for this very reason the activities must also be the same in number (since an activity that is one in number belongs to something that is one in number), but if the state is one, perhaps the activity might still not be regarded by someone as being one (since whenever one stops walking, the walking is no longer, but will be again when one is walking again). But if this is one and the same, then one and the same thing would admit of being destroyed and being many times.

20 Now these impasses are outside the present inquiry; but since every motion is continuous, then also the motion that is simply one must be continuous, if in fact every motion is divisible, and if it is continuous it is a unity. For not every motion would become continuous with every motion, just as neither would any other chance thing become continuous with a chance thing, but only those of which the extremities are one. But of some things there are not extremities, and of others the extremities are different in kind though like-named; for how could the end of a line and the end of a stroll touch and become one? Motions that are not the same in either species or genus might be next to each other (for after running, someone might immediately catch a fever), as also a torch race by relays is a consecutive change of *30* place, but not a continuous one. For the continuous is set down as that of which the extremities are one. So they are consecutive and sequential by virtue of the time's being *228b* continuous, but continuous by virtue of the motions' being so, and this is when both become one at the extremities. Hence the motion that is simply continuous and one must be the same in kind, of one thing, and in one time: in the one time, lest there be motionlessness in its midst. (For in one that leaves intervals, there must be rest; and the motion will be many and not one, of which there is rest in its midst, so if any motion is cut off at intervals by standing still, it is neither one nor continuous, and it is cut off if there is a time in between.) And of a motion that is not one in kind, even if it does not leave gaps, the time *10* indeed is one, but the motion is various in kind; for motion that is one must also be one in kind, though motion that is one in kind need not be simply one. What motion, then, is simply one, has been said; and further, what is complete is said to be one, whether in genus, in species, or in its own being, just as also with other things, completeness and wholeness belong to what is one. But there are also times when what is incomplete is called one, just so long as it is continuous.

Moreover, in another way besides those mentioned, a motion is called one if it is uniform. For the non-uniform is, as it seems, not one, but rather the uniform is one, like the straight, since the non-uniform is divided. But it seems to admit of a difference of more and *20* less. And it belongs to every motion to be uniform or not, for a thing might alter uniformly, or be carried along something uniform like a circle or straight line, and likewise with growth and shrinking. And non-uniformity is sometimes a difference in that along which a thing is moved (for it is impossible that motion be uniform that is not along a uniform magnitude, such as motion on a broken line, or on a spiral, or on another magnitude of which a random part does not fit on any random part). But sometimes the non-uniformity is not in that which moves, nor in the when, nor in the direction, but in the way it moves. For it is sometimes distinguished in quickness and slowness, and that of which the speed is the same is uniform, that of which it is not, non-uniform. Hence speed and slowness are not *30* kinds of motion, nor specific differences, because they go along with every specific difference of kind. So neither are heaviness and lightness kinds within the same thing, as of earth in *229a* relation to itself or fire to itself. So the non-uniform motion is one by being continuous, but less so, which also follows for change of place along a broken line; and the lesser is always mixed with its opposite. But if every motion that is one admits of being uniform or not, consecutive ones that are not the same in kind could not be one and continuous; for how could a composite of alteration and change of place be uniform? For it would fall short of fitting along itself.

Commentary on Book V

Books V–VII of the *Physics* get less attention than do the other books, and when they are discussed it is sometimes only for the sake of finding contradictions between things said there and in the other books. The belief in such contradictions results from misunderstanding the way dialectical inquiry works, and it can lead to pointless speculation about Aristotle's "development." It is rather the understanding of nature and of motion that develops through the *Physics,* and Book V is the clearest structural pointer to the shape and unity of the argument. Here, threads from the first three books are gathered together to begin an intense forward motion. The four preceding sections, on the infinite, place, the void, and time, argue to the local conditions and temporal implications of motion, but Aristotle here returns to an examination of motion itself.

It was said in Book I that every motion is between contrary states of some kind, in Book II that many motions are incidental to the things that move, and in Book III that all motions seem to be continuous and that they are of four kinds, change of thinghood, quality, amount, or place, while throughout Books I–III motion and change are used as synonyms. In this book, Aristotle first sets aside all incidental motions, since they are derivative and do not reveal what motion is in its own right. In its own right, motion must always be from one contrary to the other, for just the reasons given in I, 5, but close attention to this fact now reveals a surprising conclusion. There is no being that is the contrary of an independent thing, so coming-into-being is a change from a contradictory state rather than a contrary one, from what is not a human being, say, and not from something opposite to a human being. This means in turn that coming-into-being is a discontinuous change, that conception is an instantaneous event, a sudden transformation in the generative material that comes from the parents. And therefore the provisional statements about motion in the early books turn out not to cohere, and not all to be strictly true.

Change is of four kinds, but motion only of three, since only changes of quality, amount, and place are stretched out between contraries. The distinction between change and motion is not a report of the way words are used, but departs from ordinary speech. It is rooted in and confirmed by the definition of motion. That definition was formulated on the assumption that all change is motion, but it leads to a revision of its own presupposition. Motion is a being-at-work-staying-itself of a potency, and that is a holding-on in a certain state of completeness *(enteles-echein),* which is a kind of continuity or holding together *(sun-echein).* Only a change that remains one and the same through a gradual transition can fit that description of a potency at-work and staying-itself. The opposites that begin and end a motion must therefore be boundaries of an in-between range of conditions, and this is what Aristotle means by contraries, unlike contradictories with their excluded middle. The three particular kinds of motion are confirmed by comparison to the eight ultimate kinds of being, since relations, actions, and states of being-acted-upon can change only incidentally, and thinghood has no contrary. (Time was separated off at 224b, 1, as the in-which, never the from-which or to-which of a motion.) Quality, amount, and place remain, and they are the only ways of being that admit of continuous variation.

Continuity did not play a major role in the discussion of motion in Book III, but it is on the basis of that discussion that it now comes to the foreground. And even the definition of continuity given in passing in III, 1, as infinite divisibility, is unrevealing and incorrect. At 227a, 10, Aristotle redefines the continuous correctly as that of which the extremities are one. In VI, 1, he will show that infinite divisibility is a necessary but insufficient property of anything continuous. It is crucial that the extremities be one and not merely coincident, as Aristotle says at 227a, 23–24; at 211a, 31–34, he pointed out that a jar and the water in it share a common surface, but they are not continuous because their boundaries merely coincide but are not one. The unity of any motion depends obviously on the unity of the moving thing and the identity of its beginning and end states, but it depends less obviously but just as necessarily on its continuity. This will become the topic of Book VI, as Aristotle looks within motion to its internal structure.

(Chapters 5 and 6 of Book V, which explore in detail the question of which motions are contrary to which, and which states of rest are opposed to which motions, are omitted here, and removed to the Appendix.)

Book VI

Internal Structure of Motions

Chapter 1

231a, 20 If what are continuous, touching, or next in series are as they were defined earlier, continuous those things of which the extremities are one, touching those of which the extremities are coincident, and next in series those of which nothing of the same kind is between, then it is impossible for anything continuous to be made of indivisible things; for example, a line cannot be made of points, if the line is continuous and the point indivisible. For neither are the extremities of the points one (since there is not an extremity as opposed to some other part of an indivisible thing), nor are their extremities even coincident (since there is no extremity at all of that which has no part, for the extremity is different from that of which *30* it is the extremity). Yet the points of which the continuous thing is made would have to be either continuous or touching one another; and the same argument applies to everything *231b* indivisible. They could not be continuous for the reason stated, and everything touches either whole to whole, part to part, or part to whole. But since the indivisible is without parts, it could only touch whole to whole. But what is touching whole to whole would not be continuous. For the continuous has one part distinct from another and is thus divided into things that are different and separated in place. But neither could a point be next in series to a point, or a now to a now in such a way that the length or time would be made of them; for next in series are those things of which nothing of the same kind is between, but between *10* points there is always a line and between nows a time. Further, it would be divisible into indivisibles, if each thing is divisible into those things of which it is made, but it was seen that no continuous thing is divisible into things without parts. And it is not possible that another kind be in between, for it would be either indivisible or divisible, and if divisible, either into indivisibles or always into divisible things, and this is continuous. And it is clear that everything continuous is always divisible into divisible things, for if it were divided into indivisibles, an indivisible would be touching an indivisible, since the extremity of continuous things is one and touching.

And it belongs to the same argument for a magnitude, a time, and a motion to be *20* composed of and divided into indivisibles, or none of them. And this is clear from the following things. For if a magnitude is composed of indivisibles, then also the motion through it would be made of equally many indivisible motions; for example, if ABC is made of the indivisibles A, B, C, the motion DEF by which Z is moved along ABC has each part indivisible. So if when motion is present, something must be moved, and when something is moved, motion is present, then also the being-moved would be made of indivisibles. So Z was moved A in the motion D, B in E, and similarly C in F. But if what is moved from *30* somewhere to somewhere necessarily cannot at the same time be moving and have finished moving along what it moved through when it was moving (as, if something walks to Thebes, *232a* it is impossible at the same time to be walking to Thebes and have finished walking to Thebes), then Z was moved along A, which has no part, in the time in which the motion D was present; therefore, if it had passed through later than it was going through, the motion would be divisible (for when it was going through, it was neither at rest nor had it passed through, but it was in between), but if at the same time it is passing through and has passed through, then what is walking, when it is walking, will be something that has walked there and has moved to where it is moving. But if something is moved along the whole of ABC, and the motion which it moves is DEF, but it *has* moved through A, which has no part, without *being* moved through it at all, the motion would be made not of motions but of [jump-like] movements, and by way of a thing's having been moved without being in motion; for it *10* would have passed through A without going out through it. So it would be possible for something to have finished walking without ever walking, since it has finished walking this part of the magnitude without walking it. Then if everything must be either at rest or in motion, Z is at rest in each of the places A, B, and C, and so it will be possible for something to be continuously at rest and at the same time in motion. For it was moved along the whole ABC, and was at rest in every part whatever, and so also in the whole. And if the indivisibles in DEF are motions, it would be possible when a motion is present in it for a thing not to be moved but be at rest; but if they are not motions, it would be possible for a motion to be made of things other than motions. And the time must be indivisible in the same way as the length and the motion, and must be composed of indivisible nows; for if *20* every motion is divisible, but in less time something of equal speed goes through less distance, the time will also be divisible. But if the time is divisible in which something is carried along A, then A too will be divisible.

Chapter 2

Since every magnitude is divisible into magnitudes (for it has been shown that it is impossible for anything continuous to be made of uncut-able parts, and every magnitude is continuous), necessarily a faster thing will be moved more in an equal time, equally in a less time, and also more in a less time, just as some people define the faster. For let A be *30* faster than B. Since, then, what changes sooner is faster, in the time in which A has changed from C to D, say FH, B will not yet be at D in this time, but be short of it, so that, in the equal time, the faster will go through more. But surely it will go through more in a lesser time as well, for in the time in which A has come to be at D, B, being slower, will be *232b* at E. So since A has come to be at D in the whole time FH, it will be at G in a time less than this; let it be FK. So while CG, which A has gone through, is greater than CE, the time FK is less than the whole FH, so in less time it goes through more. But it is clear from these things also that the faster goes through an equal amount in less time. For since it goes through more in less time than the slower, but itself, taken by itself, goes through more in *10* more time than in less, say through LM rather than LQ, the time PR in which it goes through LM would be greater than PS, in which it goes through LQ. So if PR is less than the time X, in which the slower thing goes through LQ, PS too is less than X, since it is less than PR, and what is less than a lesser is itself also less. Therefore in a lesser time it will go through an equal amount. Further, if everything must be moved in an equal, lesser, or *20* greater time, and the slower is moved in a greater one and that with an equal speed in an equal one, while the faster is neither of an equal speed nor slower, the faster could not be moved in either an equal or a greater time. What is left is that it is moved in a lesser time, so necessarily the greater goes through the equal magnitude as well in a lesser time.

And since every motion is in time, and in every time it is possible for a thing to be moved, and every moved thing admits of being moved both faster and more slowly, then in every time the faster will also be able to be moved more slowly. These things being so, it is also necessary that time be continuous. I call continuous that which is always divisible into divisible parts, for once this is set down about the continuous, time must be continuous. For since it has been shown that the faster goes through the equal in a lesser time, let A be the *30* faster and B the slower, and let the slower have been moved through the magnitude CD in the time FH. It is clear, then, that the faster will be moved through the same magnitude in a lesser time, and let it have been so moved in the time FG. In turn, since the faster has gone through the whole CD in the time FG, the slower in the same time will go through a *233a* lesser magnitude, and let it be CK. But since B, the slower, in the time FG has gone through CK, the faster will go through it in a lesser time, so that the time FG in turn will have been divided. And when this is divided, the magnitude CK will also be divided in the same ratio. And if the magnitude also the time. And this will always be possible for those who take something away from the faster after the slower, and from the slower after the faster, and make use of the thing deduced; for the faster will divide the time, and the slower the length. If, then, it is true that one can always turn back and forth, and a division always *10* happens to that to which one turns, it is clear that every time will be continuous; for the time and the magnitude will be divided into the same and equal divisions.

And besides, it is clear that it is said on the basis of the familiar arguments that if time is continuous, so also is a magnitude, so long as a half is gone through in half the time, or simply less in less time; for there will be the same divisions of the time and of the magnitude. And if one of the two is infinite, so too is the other, and in the way the one is, *20* so too is the other, as, if the time is infinite at its ends, the length is too at its ends, but if the time is infinite by division, the length is too by division, or if the time is infinite in both ways, the magnitude is too in both ways.

And for this reason, Zeno's argument is false, since it assumes the impossibility of going through what is infinite or of touching each of infinitely many things in a finite time. For length and time, and generally every continuous thing, are called infinite in two ways, either by way of division or at the ends. Things infinite by reason of amount, then, do not admit of being touched in a finite time, but those infinite by way of division do admit of it, *30* since time itself is also infinite in that way. So it turns out that a thing goes through the infinite in an infinite, and not in a finite, time, and touches infinitely many things in infinitely many times, and not in a limited number.

And so it is not possible either to go through the infinite in a finite time or through the finite in an infinite time, but if the time is infinite, the magnitude will be infinite too, and *233b* if the magnitude, also the time. For let there be a finite magnitude AB, and an infinite time C; and let something finite, CD, be taken out of the time. In this time, then, something goes through some of the magnitude, and let what is gone through be BE. And this may either measure out AB, or fall short or long in doing so, since it makes no difference. For if something always goes through a magnitude equal to BE in an equal time, and this measures out the whole, the whole time in which it goes through it will be finite, since it and the magnitude will be divided into equally many parts. In other words, if not every magnitude is gone through in an infinite time, but some one, such as BE, admits of being gone through *10* in a finite time, and this measures the whole magnitude, and the equal is gone through in an equal time, therefore the time also will be finite. And that BE is not gone through in an infinite time, is clear if the time is taken as limited at one end; for if the part is gone through in a lesser time, this has to be finite, since one limit is there all along. The demonstration is the same if the length is assumed infinite and the time finite.

It is clear, then, from what has been said, that neither a line nor a surface nor in general any of the continuous things will be indivisible, not only because of what was said *20* just now, but also because it would turn out that one could divide the indivisible. For since in every time there is a faster and a slower, and the faster goes through more in an equal time, and it is capable of going through a length double or half-again as long (for this would be the ratio of the speeds), let the faster then be carried half-again as far in the same time, and let the magnitudes be divided, that of the faster into three indivisible parts AB, BC, and CD, and that of the slower into two, EF and FH. Accordingly, the time too will have been divided into three indivisible parts, since the equal magnitude is gone through in an equal time. So let the time be divided into KL, LM, and MN. Again, since the slower has been *30* carried through EFH, the time also will have been cut in two. Therefore the indivisible will have been divided, and that which has no part will be gone through not in an indivisible time but in a greater one. It is clear then that none of the continuous things is without parts.

Chapter 3

But it is also necessary that the now be indivisible—not what is called now as a consequence of anything else, but what is in itself and primarily so called—and such a thing *234a* must be present in every time. For there is some limit of what has happened, on this side of which there is nothing of what is going to be, and in turn a limit of what is going to be, on this side of which there is nothing of what has happened, which we claim is a boundary for both. And if this is shown to be such and to be the same thing, it will be clear at the same time that it is indivisible. But it must be the same now that is the extremity of both times, for if they are different, one could not be next in series to the other, since what is made of things without parts is not continuous, while if they are separate a time would be in between; for every continuous thing is such that there is something of the same name *10* between its limits. But surely if there is a time in between, it is divisible, since every time has been shown to be divisible. Therefore the now would be divisible. But if the now is divisible, something of the past would be in the future, and something of the future in the past; for that at which it is divided will mark off the time that has gone past, and that which is going to be. But by the same token, the now would not be so in its own right, but as a consequence of something else, since the division is not produced by it. And on top of these things, some of the now will be past and some of it future, and not always the same past or future. So neither is the now the same, for a time is divided in many places. Therefore *20* if it is impossible for these things to belong to it, it is necessary that the now in both of the two be the same. But surely if it is the same, it is clear that it is indivisible, for if it were divisible, the same things would follow which did before.

That, then, there is something indivisible in time, which we call the now, is clear from what has been said; and that nothing is moved in the now, is clear from the following. For if it were, it would admit of being moved both faster and more slowly. So let N be the now, and during it let the faster thing have been moved through AB. Then the slower in the same time will have been moved less than AB, say AC. But since the slower has moved AC *30* in the whole now, the faster will have moved that far in less than this, so that the now will have been divided. But it was shown to be indivisible. Therefore it is not possible to be moved in the now.

But neither is it possible to be at rest; for we say to be at rest that which, being of such a nature as to be moved, is not moved when, or where, or in the manner that is its nature, so since in the now nothing is of such a nature as to be moved, it is clear that neither *234b* can anything be at rest. Further, if the now is the same in both of the two times, and it is possible for something to be moved for the whole of one and at rest for the whole of the other, and that which is moved for the whole of a time will have been moved in whatever part of it in which it is of such a nature as to be moved, and what is at rest likewise will be at rest, the same thing turns out at the same time to be at rest and be in motion; for the same extremity belongs to both of the times, namely the now. Besides, we say to be at rest that which holds on in the same condition, both itself and its parts, now as before; but in the now there is no before, so there is no being-at-rest. It is necessary, therefore, both that the moving thing be moved and the resting thing be at rest in a time.

Chapter 4

10 Every changing thing is necessarily divisible. For since every change is from something to something, and when it is at that to which it changed it is no longer changing, but when it was at that from which it changed, both itself and all its parts, it was not yet changing (for that which, both itself and its parts, holds on in the same condition, is not changing), it is necessary then that something belonging to the changing thing be in the one condition and something else be in the other; for it is impossible that it be in both or in neither. By that to which it changes, I mean the first thing resulting from the change, as gray rather than black from white, since the changing thing need not be in either of the extremes. *20* It is clear then that every changing thing will be divisible.

But a motion is divisible in two ways, in one way by the time, but in another according to the motions of the parts of the thing moved; for example, if the whole AC is moved, both AB and BC will be moved. Then let DE be the motion of the part AB, and EF that of part BC. So the whole DF must be the motion of AC. For it will be moved according to this, since each of its parts is moved according to each motion; but nothing moves in the motion of anything else, so the whole of the motion is the motion of the whole *30* magnitude. And besides, since every motion belongs to something, but the whole motion DF belongs neither to either of the parts (for each of those motions is of a part), nor to anything else (for that to which the whole belongs is a whole, and the parts belong to its parts, while the parts belong to AB and BC and to nothing else, since one motion would not belong to *235a* more than one thing), the whole motion would belong to the magnitude ABC. And still, if the motion of the whole is something else, say GI, the motion of each of the parts could be subtracted from it; and these will be equal to DEF, since there is one motion of one thing. So if the whole GI is divided into the motions of the parts, GI will be equal to DF; but if anything is left as a remainder, say KI, this will be a motion belonging to nothing (for it belongs neither to the whole, nor to the parts, because the motion of one thing is one, nor does it belong to anything else, since a continuous motion belongs to what is continuous), and similarly if something exceeds GI as a result of the division; so if this is impossible, they must *10* necessarily be the same and equal. This, then, is the division according to the motions of the parts, and it must belong to everything that is divisible into parts; but that according to time is different, for since every motion is in time, and every time is divisible, and in the lesser time the motion is less, every motion must be divisible according to time. And since everything moved is moved in some respect and for some time, and of every one of them there is a motion, there must be the same divisions of the time, the motion, the moving, the thing moved, and that in respect to which the motion is (except not in the same way of all the things in respect to which there is motion, but of place in its own *20* right, of quality incidentally). For let a time A be taken in which something is moved, and let the motion be B. If, then, the thing has been moved through the whole motion in the whole time, then it has been moved less in the half time, and less again than that when this has been divided, and in the same way forever. And similarly, if the motion is divided, the time is divided, for if it is the whole motion that has been moved in the whole time, it is the half in the half, and again the less in the lesser. And in the same way the moving will be divided, for let the moving be C. As a result of the half motion there will be less than the whole of the moving, and again as a result of the half of the half, and so on forever. And it is possible for the moving resulting from each of the motions to be set out, such as those *30* resulting from DC and CE, in order to claim that the whole will be resultant from the whole (for if something else were, there would be more than one moving resulting from the same motion), just as we also showed that the motion is divided into the motions of the parts; for once the moving resulting from each motion has been taken, the whole will be continuous. Likewise, it will also have been shown that the length is divisible, and in general everything in respect to which there is change (except that some are so incidentally, because the changing thing is divisible); for when one of these things has been divided, all of them will *235b* have been divided. And as for their being finite or infinite, the same thing will hold for all of them. But most of all, the dividing and being infinite of them all follows upon the changing thing, for it is to the changing thing that divisibility and infinity immediately belong. Divisibility has been shown above, and infinity will be clear in what follows.

Chapter 5

Since every changing thing changes from something to something, what has changed, when it has first changed, must be in the condition to which it has changed. For *10* what changes something stands apart from that from which it changes, or leaves it behind, and either the changing and the leaving behind are the same, or the leaving behind follows upon the changing. But if leaving behind follows upon changing, having left something behind follows upon having changed, for each is related similarly to each. Then since one of the changes is in respect of a contradictory, when something has changed from not-being to being, it has left not-being behind. Therefore it will be in the condition of being, since everything must either be or not be. So it is clear that in a change with respect to a contradictory, what has changed will be in the condition to which it has changed. But if it is so in this case, it is so also in the others, for it is the same way with one as with the others. And this will be still *20* more clear to those who take up each case one by one, if indeed what has changed must be somewhere or in some condition. For since it has left behind that from which it has changed, and it has to be somewhere, it will be either in that place or condition to which it changed or in some other. But if it is in another, say that which has changed to B is in C, it is changing in turn from C to B; for it was not assumed to be following after B, and change is continuous. Therefore the thing that has changed, when it has changed, is changing into that to which it has changed, which is impossible; therefore it is necessary for what has changed to be in that to which it has changed. It is clear, then, also that what has come into being will be, and what has been destroyed will not be; for the conclusion was *30* stated generally about every change, and it is most evident in change with respect to a contradictory.

That, then, what has changed, when it has first changed, is in that condition, is clear; and that time in which it has first changed is necessarily indivisible. I call "first" that which is such-and-such not because anything other than it is so. For let AC be divisible, and let it be divided at B. If, then, something has changed in AB or else in BC, it would not have *236a* changed first in AC. But if it changed in each of them (since it must either have changed or be changing in each), it would be changing in the whole time; but it was assumed to be something that had changed. And the argument is the same if it is changing in one part and has changed in the other; for there will be something before what is first. So that in which it has changed could not be divisible. It is clear then that what has been destroyed has been destroyed, and what has come into being has come into being, in an indivisible time.

But the time in which something has first changed is meant in two ways: that in *10* which the change was first completed (for then it is true to say that it has changed), or that in which it first began to change. Now the one that is meant in reference to the first completion of the change belongs to it and exists (for a change admits of being completed and there is an end of change, which was shown to be indivisible on account of being a limit); but the one that refers to the beginning does not exist at all, since there is no beginning of change, nor any portion of time in which it first changed. For let the first time be AD. Now this is not indivisible, since it would follow that the nows were next to each other. And further, if it is at rest in the whole time CA (for let it be set down as being at rest), then it is also at rest in A, so if AD has no parts, it will at the same time be at rest and *20* have changed, since it is at rest in A and has changed in D. But since it is not without parts, then necessarily it is divisible and the thing has been changing in whatever part of it one might take; for if AD is divided, and if the thing has changed in neither part, then it has not changed in the whole either, while if it changes in both it does so also in the whole, but if it has changed in one of the two, it has not changed *first* in the whole. So it must have been changing in every part whatever. It is clear, then, that there is no time in which a thing has first changed, since the divisions are infinite.

Nor of the thing that has changed is there any first part which has changed. For let *30* DF be the first part that has changed of DE, since every changing thing has been shown to be divisible, and let the time in which DF has changed be GI. If, then, it has changed in the whole of DF, some smaller part will have changed in half the time and before DF, and another in turn before this, and yet another before that, and so on forever. So there will be no part of the changing thing that has changed first.

That, then, there is no part of the changing thing nor time in which it changes that is *236b* first, is clear from what has been said; but it is no longer the same story with that itself which changes, or according to which the thing changes. For there are three things which are spoken of in a change: the changing thing, that in which it changes, and that to which it changes, say a human being, the time, and whiteness. The human being, then, and the time, are divisible, but about the whiteness it is a different story. Yet there is an exception even here, since incidentally everything is divisible, for that of which whiteness or any quality is an attribute is divisible, and since whatever is called divisible in its own right and not *10* incidentally will not have in it any first part, just as there is none in magnitudes. For let there be the magnitude AB, and let it have been moved from B to C first. Then if BC were indivisible, that which has no part would be next to that which has no part; but if it is divisible, there would be something before C, to which it has changed, and before that in turn something else, and so on forever, since the division never gives out. So there will be no first thing to which it has changed. And it is the same with change of quantity, for this too is in respect to something continuous. It is clear, then, that in motion with respect to quality alone is it possible for there to be anything indivisible in its own right.

Chapter 6

20 And since every changing thing changes in time, but is said to change in a time not only in the sense of the one in which it primarily is, but also in the sense of one in which it is as a result of something else, as in a certain year because it changes in a certain day, it is necessary that the changing thing be changing in every part whatever of the time in which it primarily is. This is clear even from the definition (for primary is meant in this way), but nonetheless it is clear too from what follows. For let XR be the primary time in which a moving thing is moved, and let it be divided at K, since every time is divisible. So in the time XK it either moved or did not move, and likewise in turn in KR. But if it were moved in neither, it would have been at rest in the whole (for it is impossible to be moved *30* in none of the parts of the time it was moved); but if it were moved in one of the two only, it would not have been moved in XR primarily, since the motion would be in it as a result of something else. It is necessary, therefore, that the thing be moved in every part whatever of XR.

Now that this has been shown, it is clear that everything that is moving has been moving earlier. For if in the primary time XR, something has been moved through the magnitude KL, then in half the time, something with the same speed which started at the *237a* same time will have moved through half of it. But if the thing with equal speed has been moved a certain distance in the same time, the other one too must have moved the same magnitude, and so the moving thing will have been moving. And further, if we say something has been moved in the whole time XR, or in any time in general, by coming to the last now of it (since this is the boundary, and what is between nows is time), then also in the same way it should be said to have moved in the other times. And the division is the extremity of its half. So it will have been moved also in the half time, and in general in any *10* of the parts whatever, since by means of the cut, every time is always bounded by the now there with it. Then if every time is divisible, and what is between the nows is a time, every changing thing will have undergone an infinity of changes. And further, if what is changing continuously, and has not been destroyed or left off from the change, must necessarily either be changing or have changed in every part whatever of the time, but is not able to be changing in the now, it must have changed at each of the nows; so if the nows are infinite, every changing thing has undergone an infinity of changes.

And not only must the changing thing have changed, but also the thing that has changed must at an earlier time be changing, since everything that has changed from *20* something to something has changed in time. For let there be something that has changed from A to B in the now. Now it has not changed in the same now in which it is in condition A (for it would be at the same time in both A and B); for that the thing that has changed, when it has changed, is not in that from which it has changed, has been shown above. But if it is in another now that it has changed, there will be a time between them, since nows are not next to each other. Then since it has changed in time, and every time is divisible, in half the time there will have been another change, and in half of that another in turn, and so on forever; so it would be changing beforehand.

And what was said is still clearer with the magnitude, on account of the continuity of *30* the magnitude along which the changing thing is changing. For let there be something that has changed from C to D. Then if CD is indivisible, that which has no part will be next to that which has no part; and since this is impossible, what is between them must be a magnitude and be infinitely divisible, so that it changes into that infinitude of things beforehand. *237b* Therefore necessarily everything that has changed is changing beforehand. For the same demonstration holds even among what is not continuous, not only in change between contraries but also in change to a contradictory; for we shall take the time in which the thing has changed, and say the same things again. So necessarily what has changed must change and what changes must have changed, and the having changed is before the changing and the changing before the having changed, and never will we have come to a first. And the cause of this is that what has no part is not next to what has no part; for the division is infinite, just as with extensions and divisions of lines.

10 It is clear, then, that also what has come into being must be coming into being beforehand, and what is coming into being must have come into being, so long as they are divisible and continuous; what has come into being is not, however, always the very thing which is coming into being, but sometimes something else, as in the case of those things of which it is some part, for example the foundation of the house. And similarly with what is being destroyed and what has been destroyed: for there is immediately present in what comes into being or is destroyed, something infinite, since they are continuous, and it is not possible either to have come into being without something coming into being or to be coming into being without something having come into being, and similarly with what is being destroyed and what has been destroyed. For always there will be that which has been destroyed before the being destroyed and what is being destroyed before the having been destroyed. Then it is clear *20* also that what has come into being must be coming into being beforehand, and what is coming into being must have come into being; for every magnitude and every time is always divisible. Therefore in that along which or during which it is, it could not be in any first.

Chapter 7

Since every moving thing is moved in time, and through a greater magnitude in more time, it is impossible for a thing to be moved a finite magnitude in an infinite time, if it is not repeatedly moving through the same magnitude and some of it during each time, but along all of it for all the time. That, then, if anything should be moved at a constant *30* speed, it must be moved the finite magnitude in a finite time, is clear. (For if a part is taken which measures off the whole magnitude, it will have been moved through the whole in as many equal times as there are parts, so since these parts are finite, both in the size of each and the number of them all, the time too would be finite; for so many of them will amount to so much as the time of each part multiplied by the number of the parts.) And even if it *238a* is not at a constant speed, it makes no difference. For let there be the finite interval AB, which is moved through in an infinite time, and let the infinite time be CD. Now if the thing must have moved through one part before another (and this is clear, that it has moved through different parts in the earlier and later parts of the time, since it will always have moved through something else in a greater time, whether it changes at a constant speed or not, and nonetheless whether the motion intensifies or slackens or stays the same), let some part of the interval AB be taken, namely AE, which measures off AB. Now this happened in some time marked out of the infinite one; it could not be during the infinite one, since the whole *10* interval is moved through in the infinite one. And so if I again take as much as AE, it must be moved through in a finite time, since the whole is in the infinite one. And continuing to take parts in this way, since there is no part of the infinite which will measure it off (for it is impossible that the infinite be made out of finite things, equal or unequal, since it is things finite in magnitude and multitude that will have been measured off by some one part, and no less so whether they themselves are equal or unequal, but definite in magnitude), while the finite interval is measured by a certain number of parts AE, then AB would have been moved through in a finite time (and similarly with a process of coming to rest). And so it is not possible for anything that is one and the same to be coming into being, or passing away, forever.

20 And the argument is the same that it is not possible for anything either to be moved or to be coming to rest over an infinite magnitude in a finite time, neither moving uniformly nor nonuniformly. For if some part is taken which will measure out the whole time, in this time the thing goes through a certain amount of the magnitude and not the whole of it (since it goes through the whole in the whole time), and another part again in an equal time, and similarly in each one, whether equal or unequal to the one at the beginning, since it makes no difference so long as each is finite; and it is clear that, when the time has been used up, the infinite magnitude will not have been used up, since the process of subtraction *30* becomes finite, both in how much is taken away and how many times. Therefore it does not go through the infinite in a finite time. And it makes no difference whether the magnitude is infinite in one direction or in both, since the argument will be the same.

Now that these things have been demonstrated, it is clear that neither will a finite magnitude be able to go through an infinite in a finite time, for the same reason; for in part of the time it will go through a finite amount, and likewise in each part, and therefore a *238b* finite amount in the whole time. And since the finite magnitude does not go through the infinite one in a finite time, it is clear that neither does an infinite one go through a finite; for if the infinite goes through the finite, then necessarily the finite also goes through the infinite. For it makes no difference which of the two is the one moved, since both ways the finite goes through an infinite. For whenever the infinite A is moved, some of it will be along the finite B, say CD, and again another part and another, and so on always. So at the *10* same time it will follow that the infinite has moved the finite magnitude and the finite has gone through the infinite; for neither is it even possible, perhaps, for the infinite to be moved through the finite otherwise than by the finite going through the infinite, either as a thing carried along it or one which measures it off. So since this is impossible, the infinite could not go through the finite. But surely neither does the infinite go through the infinite in a finite time; for if it goes through the infinite, then it also goes through the finite, since the finite is present in the infinite. Besides, if one takes part of the time, the demonstration will be the same.

And since the finite does not move through the infinite, nor the infinite through the *20* finite, nor the infinite through the infinite in a finite time, it is clear that neither will there be an infinite motion in a finite time. For what difference does it make whether the motion or the magnitude is made infinite? For necessarily, if either one is infinite, the other is too, since every change of place is in a place.

Chapter 8

Since everything that is of such a nature as to move or be at rest does so when, where, and how it is natural to it, what is stopping must be moving when it is stopping; for if it is not moving it will be at rest, but it is not possible for what is at rest to be coming to rest. Now that this has been demonstrated, it is clear that it is also necessary that a thing come to a stop in a time (since the moving thing moves in time, and the thing 30 that is stopping has been shown to be moving, so that it must be stopping in time); and what's more, if we speak of a faster and a slower in time, there is such a thing as stopping faster and more slowly.

And in every part whatever of that time in which the stopping thing is primarily stopping, it must be stopping. For if the time is divided, and if it is stopping in neither of the parts, then neither is it doing so in the whole, so that the stopping thing would not be stopping. But if it is stopping in one of them, it would not be stopping in the whole time primarily, since it is stopping in this as a result of something else, just as was said above *239a* about the moving thing. And just as the moving thing is not moving in any first time, in the same way, neither is the stopping thing stopping in any first, for neither of moving nor of stopping is there any first. For let AB be the time in which it is first stopping. Now this cannot possibly be without parts (for there is no motion in a time that has no part, since something has moved before it [see 237a, 11–15], and what is stopping has been shown to be moving), but surely if it is divisible, it is stopping in any of the parts of it whatever, since it has been shown above that in the time in which it is primarily stopping, it is stopping in every part of *10* it whatever. Since then it is a time in which it is stopping primarily, and not something indivisible, since every time is infinitely divisible, there will not be any first time in which it is stopping.

But neither is there any first time at which the resting thing is at rest. For it does not rest in a time which has no parts, since motion is impossible in an indivisible time, while in that in which there is resting, there is moving also (for we said that a thing is resting when it is by nature not moving at a time when it is of such a nature as to be moving); and besides, we say that something is resting whenever it is in the same condition now as before, as though judging it not by any one thing but by two at least, so that it will not be possible that it be at rest in that which has no parts. But if it has parts, it will be a time, and the thing will be at rest in any of its parts whatever. For this could be shown in the same way *20* as with the things above, and so there will be no first time. And the cause of this is that everything rests and moves in time, and it is not possible for there to be a first time, or magnitude, or anything continuous at all, since all of them are divisible into an infinity of parts.

And since every moving thing moves in time and changes from something to something, at a time in which it is moving in its own right and not by means of what is in some part of it, it is impossible that the moving thing be primarily at any place or condition at that time. For resting is being in the same condition for some time, both a thing itself and each of its parts. For this is how we speak of being at rest, whenever in one and *30* another of the nows it is true to say that a thing itself and its parts are in the same condition. But if this is resting, the changing thing cannot be as a whole at any condition, at the primary time it is changing; for every time is divisible, and so at one part and another of it, it would be true to say that the thing itself and its parts were in the same condition. For if this were not so, but at only one of the nows was the thing in this condition, it would not be at it for any time, but at the limit of the time. And in the now there *is* always some *239b* condition at which it is, but it is not at rest, for neither moving nor resting is in the now, so in spite of its being true that it is not moving in the now, and is at some condition, still in time it does not admit of being at rest in any condition, since it would follow that the thing being carried along was at rest.

Chapter 9

Zeno's reasoning, then, misses the mark: for he says that if everything is always at rest when it is at a place equal to it, while what is changing place is always doing so in the now, the flying arrow is motionless. But this is false, for time is not composed of indivisible *10* nows, just as no other magnitude is. And there are four arguments of Zeno about motion, which give indigestion to those who unravel them; the first is the one about there being no motion since the moving thing must have come to the half before the end, which *we* have gone all the way through in the passage above [232b, 20–233a, 32]. Second is the so-called Achilles, and it is this: that the slowest, running, will never be left behind by the fastest, since before that the pursuer must have come to the place the pursued set off from, so that the slower is necessarily always in front by some amount. But this is the same argument as the bisecting, *20* but differs in that the dividing of the magnitude taken beforehand is not in half. The not catching up with the slower follows from the argument, but it comes about by means of the same thing as the bisecting (for in both, the magnitude being divided in some way, the not coming to the end follows, but in this one it is piled on that not even will the one represented as fastest in the tragedies do so in pursuing the slowest), so the refutation too must be the same. And holding that the one in front will not be caught is false; for when he is in front he will not be caught, but he will still be caught as long as one grants that he *30* goes through a finite distance. These, then, are two arguments, and the third is the one mentioned, that the flying arrow stands still. And it follows from taking the time to be composed of nows; if this is not granted there will not be a syllogism.

And the fourth is the one about the things of equal size which are moved opposite ways in a stadium, past equal things, some from the ends of the stadium and others from the *240a* middle, at equal speeds, in which he thinks it follows that the half time is equal to its own double. But the fallacy is in holding that something will be carried in an equal time past one equal magnitude moving at the same speed and past another equal magnitude at rest, which is false. For example, let there be placed a row of stationary equal bodies each marked A, then a row marked B beginning from their middle, being equal to these in number and magnitude, and then a row marked C from the end [of row B, that is, from the middle of row A, but extending in the opposite direction], equal again in number and magnitude and equal in speed to row B. Now it follows that when these have moved past *10* each other, the first B will be at the end at the same time as the first C. And it follows that this first C will have gone past the whole length [of row B], but the first B past the half length [of row A], so that its time will be half, since the time for each to pass each body is equal. But at the same time it follows that the first B has gone past all of row C, for the first C and the first B will be at opposite ends simultaneously, since both come to be past row A in an equal time. So this is the argument, but it follows from what was said to be false.

20 Nor will there be anything impossible for us in a change to a contradictory, as when something is changing from not-white and is in neither condition, on the ground that it is therefore neither white nor not-white. For it is not because something is not as a whole in either condition that it would not be called white or not-white, since we call something white or not-white not through its being wholly so but because most or the most important of its parts are so, and not to be in a certain condition is not the same thing as not to be in this condition wholly. And it is the same with being and not-being and the rest of the changes to a contradictory, since the thing will necessarily be in one or the other of the opposites, but all the time is in neither wholly. And it is the same again with the circle and the sphere, *30* and in general with things that are moved within themselves, because it will follow that they are at rest, since both they themselves and their parts will be in the same place for some *240b* time, and so will be at the same time at rest and in motion. For, first of all, the parts are not in the same place for any time, and next, the whole is always changing to a different condition; for the circumference taken from A is not the same as that from B or from C or from any of the other points, except incidentally, as the educated human being and the human being are the same. So one thing is always changing into another, and it is never at rest, and it is the same way with the sphere and with the other things that are moved within themselves.

Chapter 10

Now that these things have been demonstrated, we say that what has no parts *10* is incapable of moving except incidentally, as when the body or magnitude to which it belongs is moved, as is the case if something in the ship is moved by the change of place of the ship, or a part by the motion of the whole. (By what has no parts I mean what is indivisible in quantity.) For there are different motions of the parts, those belonging to the parts themselves as well as those resulting from the motion of the whole. And one might see the difference best in a sphere, since the speed of the parts near the center is not the same as that of those outside, or of the whole, as though they did not belong to one motion. Just as we said, then, what has no parts admits of moving in the way someone sitting in a *20* boat does when the boat is running, but does not admit of moving in its own right. For let it change from AB to BC, whether from a magnitude to a magnitude, from a form to a form, or to a contradictory, and let the time in which it is primarily changing be D. Necessarily, then, during the very time in which it is changing, it is in AB, or in BC, or some of it is in one and some in the other, for this held true of every changing thing. Now there will not be some of it in each, for then it would have parts. But surely neither is it in BC, for then it will have changed, but it is assumed to be changing. What is left, then, is that it is in AB *30* during the very time in which it is changing. Therefore it will be at rest, for being in the same condition for some time is being at rest. So what has no parts does not admit of moving or of changing at all, for there is only one way in which there would be motion of *241a* it, and that is if time were made of nows; for it would always have moved and have changed in the now, so as never to be moving, but always to have moved. But that this is impossible has been shown before, for neither is time made of nows, nor a line of points, nor a motion of [jump-like] movements; for this way of speaking does nothing but make the motion out of things that have no parts, exactly as if time were made of nows or a length of points.

And further, it is clear from the following that neither a point nor any other indivisible thing admits of being moved. For every moving thing is incapable of moving a *10* distance greater than itself before either an equal or a lesser one. But if this is so, it is clear that the point too will have first been moved an equal or lesser distance. But since it is indivisible, it is impossible for it to have been moved a lesser one beforehand; therefore it will have moved a distance equal to itself. And so the line will be made of points, for having always been moved an equal distance, the point will measure out the whole line. But if this is impossible, then it is also impossible that the indivisible thing be moved.

And besides, if everything moves in time, and nothing in the now, and every time is divisible, there would be for any moving thing whatever a time less than that in which it moves a distance as much as itself. For that in which it is moved will be a time since *20* everything is moved in time, and that every time is divisible has been shown above. Therefore if a point is moved, there will be some time less than that in which it has moved its own length. But this is impossible, since in the lesser time it must be moved a lesser distance. And so the indivisible would be divisible into the lesser, just as the time into the lesser time. For in only one way could that which is without parts and indivisible be moved: if it were possible to be moved in the indivisible now. For being moved in the now and the being moved of something indivisible belong to the same account.

And no change whatever is infinite; for every one was understood to be from something to something, those between contradictories as well as those between contraries. *30* So in those between contradictories, the assertion or the denial is the limit (as being of a coming-into-being, or not-being of a destruction), and of those between contraries, the contraries are, since these are extremities of change, and so they are also limits of every alteration (since alteration is according to some contraries), and likewise with increase and *241b* decrease, since the limit of growth is that of the complete magnitude in accord with the nature proper to a thing, and of decrease it is the loss of this. But change of place will not be limited in *this* way, since not every change of place is between contraries. But since what is impossible to cut, in the sense of not admitting of being cut (for "impossible" is meant in more than one way), does not admit of being in the process of being cut, neither in general can that which is impossible to have come about be coming about, nor could that which is impossible to have changed admit of being in the process of changing to that to which it is impossible for it to have changed. If, then, the thing changing place is changing to *10* something, it will also be possible for it to have changed. Therefore the motion is not infinite, nor will it have been carried an infinite distance, since it is impossible to go through it. That, then, there is not a change infinite in such a way as not to be bounded by limits, is clear. But whether any, while being one and the same change, admits of being infinite by way of time, is something that must be examined [see Book VIII, Chapter 8]. If the change is not one, nothing perhaps prevents it, as when after a change of place there should be an alteration, and after the alteration an increase, and again a coming-into-being, for in this way there would be motion forever in time, but *20* it will not be one since there is not one change made out of them all. In order to be one, a change does not admit of being infinite in time, with one exception, and this is rotation in a circle.

Commentary on Book VI

Book VI contains the most demanding sequence of intricate argument in the *Physics*. It is sometimes considered a new approach to understanding motion, independent of or even in conflict with the definition of motion in Book III. But Book VI begins from the conclusion of Book V, that all motion is continuous, which in turn is derived as a necessary consequence of the Book III definition, which itself is rooted in the argument of Book I that all change presupposes a being with a potency for something it lacks. The *Physics* is woven from a number of threads, but it is a compact and unified whole.

It is also sometimes said that the approach to motion in Book VI is of a mathematical character, but this opinion comes only from a first glance at pages full of single letters, and is true in only a limited and superficial sense. Aristotle does focus here on certain attributes of bodies and motions while ignoring most of their properties, and this is the procedure of mathematicians. It is this separate consideration of general and common attributes that permits the use of letters as labels, but Aristotle's use of them does not constitute the symbolic procedure used in algebra, in which the letters are treated as general magnitudes and operated on as though they were numbers. When Aristotle speaks of a motion A along a magnitude B in a time C, these always signify some particular motion, magnitude, and time, the identity of which is irrelevant to the argument and unspecified. This is a way of thinking about things that happen in the natural, perceptible world, and has nothing in common with the approach of Descartes, Galileo, and Newton, which is to substitute mathematical objects for the things in the world, and reason only from within mathematics. That approach of modern physics automatically eliminates any kind of motion other than change of position, but it is of the utmost importance to realize that, in almost every argument in Book VI, and unless the contrary is explicitly stated, qualitative motion is included just as much as is change of place. A motion from A to B can be a change from one state or condition to another as easily as from one place to another. In the sense that its arguments are not restricted to quantities, Book VI is not mathematical at all.

Aristotle's tentative definition of continuity in Book III, as infinite divisibility, is now seen to be a consequence of the new definition that replaced it in Book V, that the continuous is that of which the extremities are one. (This is precisely equivalent to the definition of continuity given in the nineteenth century by Richard Dedekind.) Since motion must, by its definition in Book III, be continuous, it follows that it can have no indivisible parts, and from that it follows that neither can time nor a magnitude have any indivisible parts. Aristotle does not adopt the mathematical study of geometrical continuity as a guide to understanding motion, but reasons in exactly the opposite direction. And because times and magnitudes are both divided in the same way that motions are, they are divided in the same way as one another. This conclusion is already sufficient to refute those of Zeno's paradoxes that appeal to the infinite divisibility of length. Motion is impossible, Zeno said, since a moving thing would have to traverse infinitely many places, which would take an infinite time; but the time is infinitely divisible in exactly the way the length is, and there is enough of the former to match up with all the parts of the latter. But this is an argument about parts, not about points or nows. The now is not a part of time, but an indivisible limit of a time, so it makes no sense to speak of any motion, or even of rest, as taking place in the now.

Here a surprising conclusion is reached, which again shows that Aristotle is not speaking at all about mathematical objects. Since every part of a motion and of its time is of some magnitude, no moving thing can be indivisible. We can of course imagine a point changing position, but this bears no relation to, and reveals nothing about, something that is changing color, and must, while changing, not be altogether of either the initial color it abandons or the final color it reaches. This state is intelligible only if there is divisibility in the thing, so that it can be and not be in opposite conditions in different respects. The change takes hold of the thing gradually, but this leads to another surprising conclusion: there is no first part of the thing that changes since, once it is in motion, to however small an extent, an infinitely divisible stretch of motion already precedes it. From this it follows that there is no first instant of change, nor any first position at which a body changing place is in motion.

In Chapters 5 and 6, where these last conclusions are reached, Aristotle points out that they apply not only to those changes that are between contraries, but also to coming-into-being and destruction, which are discontinuous transitions to contradictory states. But this distinction, which narrowed the most proper sense of motion from four to three kinds of change, is now extended to qualitative change as well as to change of thinghood. Where the latter is simple coming-into-being, the former is particular coming-into-being. A white thing that turns black, for instance, is becoming not-white, and once it is changing it has already passed into that contradictory state, and can no longer be said to be wholly or simply white. That transition is discontinuous, and by calling attention to this, Aristotle now effectively restricts the proper meaning of motion to only two kinds of change—change of amount and change of place. This exploration of the meaning of continuity has forced a second revision of the preliminary list of kinds of motion to include only those two that are continuous in all respects.

Despite the fact that all motion is infinitely divisible, no motion traversing a finite distance or passing between two definite conditions can take an infinite time, nor can any motion that takes a finite time cover any infinite magnitude or separation. A definite motion contains nothing of infinite extent, nor can any aspect of it be contracted to the infinitely small. In the now the moving thing is somewhere, or in some condition, but only incidentally, as passing through it, and not in the proper sense in which *being* there would mean being at rest. This brings Aristotle back to Zeno's most fundamental paradox, according to which a flying arrow is not moving, since at any time it is in one definite place.

Zeno forces us to assume, incorrectly, that time is composed of nows. (An arrow that was *at* some position in its flight, rather than moving through it, would fall straight down to earth.) The four most famous paradoxes of Zeno build up from the arrow, a moving thing considered on its own, to the bisection, Achilles, and stadium paradoxes. The bisection argument is the familiar claim that something could not move to the wall without first reaching the half-way point, after which it would have to get half-way to the wall again, and so on, so that no motion could ever be completed. But Aristotle reminds us that the refutation of this argument was completed in Chapter 2, since the finite time in which the motion is completed is divided up just like the distance. But the content of Chapter 9 has shed new light on Zeno's mistake. As with the arrow, he is assuming that time is composed of nows, which must be added together to reach the end of a stretch of time. The whole motion with its continuity is the primary datum of experience, and points and nows are derivative from it.

The Achilles paradox, as Aristotle says, contains nothing new, except in substituting Hector for the wall and setting him in tortoise-like motion. Achilles keeps having to reach positions from which Hector has already departed, but specify how far Hector goes in how much time, and Achilles will catch him, traversing infinitely many prior positions of Hector in some infinitely divisible time. The stadium paradox builds up to comparing two comparisons of motions, but obviously equivocates on the meaning of the time it takes to pass two things of equal length, since one stands still and the other has its own motion. In Chapter 2 Aristotle gave a technical refutation of Zeno's most effective arguments. In Chapter 9 he makes clear what the underlying mistakes in Zeno's assumptions are. But in a larger sense, Book VI as a whole suggests that Zeno is equivocating on motion and its possible mathematical representation, though even within mathematics he is making the mistake of supposing a line to be made of points. Aristotle's reasoning is rooted in those things that strike us in our original experience of the world. Continuity as analyzed here is first of all continuity as presented to us in those experiences in which motions endure and remain the same. That ties them to beings that endure and remain themselves despite changing, and that means that continuity of motion presupposes potencies of beings. Aristotle will address this positive side of the rejection of Zeno's arguments in VIII, 8, but first he turns in Book VII to the dependence of motions on things that cause them.

Book VII

The Relation of Mover and Moved

Chapter 1

241b, 34 Every moving thing must be moved by something. For if it does not have the source of its motion in itself, it is clear that it is moved by something else (for another *40* thing will be the mover); but if it is in itself, let AB be taken, which is moving in its own right, and not by means of some part of it being moved. First, then, to suppose that AB is moved by itself since the whole is moved and by nothing outside it,would be as if, when KL was moving LM and was itself moving, one were to say that KM was not moved by anything, *242a, 35* on account of its not being evident what was the mover and what the moved. Then too, what is not moved by anything need not stop moving because something else is at rest, but if something is at rest because something else has stopped moving, it must be moved by something. If this is taken as given, then every moving thing will be moved by something. *40* For since AB has been assumed to be moving, it must be divisible, since every moving thing is divisible. Then let it have been divided at C. Then if CB is not moving, AB will not be moving, for if it were moving, it is clear that AC would be moving while CB was at rest. But we have agreed that what is at rest when something is not moving is moved by something, so that necessarily every moving thing is moved by something, since the moving thing will always be divisible, and when the part is not moving the whole too must be at rest.

50 And since every moving thing must be moved by something, if something is moved with a motion in place by another moving thing, and the thing moving it is moved in turn by another moving thing, and that one by another, and so on always, it is necessary that there be some first mover, and that it not go to infinity; for let it not be so, and let it become infinite. So let A be moved by B, B by C, C by D, and always the next by a next. Then since it is set down that the mover causes motion by moving, the motions of the mover *60* and the moved must happen at the same time (since at the same time the mover moves and the moved is moved); it is clear then that at the same time there will be motion of A and of B and of C and of each of the things causing motion and being moved. Then let the motion of each be taken, and let that of A be E, that of B be F, and those of C and D be G and H. For though each is always moved by another, still there will be a motion of each to be taken that is one in number; for every motion is from something to something, and not infinite at its extremities. And by a motion one in number, I mean one that takes place from something that is numerically the same, to something numerically the same, in a time that is numerically the same. For it is possible for a motion to be the same in genus, in *242b, 35* species, or in number: in genus, those in the same category of being, such as changes of thinghood or of quality; in species, those from something the same in kind to something the same in kind, such as from white to black or from good to bad of a non-differing kind; but in number, that *40* from something one in number to something one in number, in the same time, such as from this white to this black, or from this place to that one, in this time, since if it were in a different time, it would no longer be a motion one in number, but one in species. But these things were spoken about in what went before [V, 4].

And let the time be taken in which A has been moved through its motion, and let it be K. And since the motion of A is finite, the time too will be finite. But since the things moving and being moved are infinite, the motion EFGH of them all will also be infinite. For *50* it is possible that the motion of A and that of B are equal, and that the motions of the other things are equal to theirs, or it is possible that the motions of the other things are greater, and so, whether they are equal or greater, in both cases the whole will be infinite; for we are assuming something possible. But since A and each of the other things are moved at the same time, the whole motion will be in the same time as that of A, and that of A is in a finite time, so that there would be an infinite amount of motion in a finite time, and this is impossible.

In this way, then, the original proposition would seem to have been demonstrated, but in fact it is not proven since nothing impossible has been shown; for it is possible that there be infinite motion in a finite time, not of one thing but of many. And this very thing happens in this case, for each thing is moved through its own motion, and that many things *60* be moved at the same time is not impossible. But if that which primarily causes motion by a bodily motion in place must either touch or be continuous with the thing moved, just as we see in every case, it is necessary that the things moved and moving be continuous with or touching one another, so as to be some one thing made of them all. And whether this is finite or infinite makes no difference to the matters now at hand, for in every case the motion, being of infinitely many things, will be infinite, so long as it is possible that the motions be either equal or greater than that of A, for we shall take what is possible as being present. If, then, the thing made of ABCD is something either finite or infinite, and is *70* moved through the motion of EFGH in the time K, which is finite, it follows that in a finite time, a thing either finite or infinite has gone through something infinite, and in both cases this is impossible. Therefore it is necessary to stop, and that there be some first mover and moved thing. For it makes no difference that the impossible thing follows from a hypothesis, *243a, 30* since the hypothesis has been admitted to be possible, and once one has set down something possible, it does not belong to it to have anything impossible come about by way of it.

Chapter 2

A first mover, not as that for the sake of which but that from which the source of the motion is, is together with the thing moved (and by together, I mean that nothing is between them), for this is shared by every mover and moved thing. And since there are three motions, of place, of the of-this-sort, and of the so-much, the movers too must be three: what carries something along, what alters something, and what increases or *40* decreases something. First, then, let us speak about change of place, for this is the first of the motions.

11 Now everything that changes place is moved either by itself or by something else. In as many of these as are moved by themselves, it is obvious that the moved and mover are together, since the first mover is present in them, so that nothing is in-between. But as many as are moved by other things must come about in four ways, for there are four kinds of change of place by means of something else: pulling, pushing, carrying, and whirling. For all motions with respect to place lead back to these; for pushing on is a kind of pushing, in *20* which that which moves something away from itself pushes while following along, whereas pushing off is one in which it does not follow along when it has moved something, and *243b* throwing is one in which it makes the motion away from itself more violent than the change of place the thing has by nature, and the thing is carried along for just so long as this motion prevails. Again, tearing apart and pressing together are pushing off and pulling: tearing apart is pushing off (since it is a pushing off either away from the mover itself or away from something else), and pressing together is pulling (since it is a pulling either toward the mover itself or toward something else). And it is the same as well with as many kinds as there are of these, such as smoothing down and combing, for the former is a pressing together and the latter a tearing apart. And it is similar with the other combinings and separatings—for they *10* will all be tearings apart and pressings together—except for those in a coming into being or a destruction. At the same time, it is clear that combining and separating are not a distinct class of motion, since all of them are divided up into some of those mentioned. Further, inhaling is pulling and exhaling pushing. And it is similar with spitting and as many other motions as there are throughout the body of secretion or absorption, for some are pulling and others pushing away. And it is necessary that the other changes of place also be led back, since they all fall into these four kinds.

And of these, in turn, carrying and whirling lead back into pulling and pushing. For *20* carrying is always a result of some one of these three ways (for the thing carried is moved *244a* incidentally, because it is in or on something that is moved, while the thing that carries it carries something either pulled or pushed or whirled, so that carrying is common to all three); but whirling is composed of pulling and pushing, for that which whirls something must pull part of it and push part, since it draws part of it away from itself and part toward itself. Therefore, if what pushes and what pulls are together with what is pushed and what is pulled, it is clear that nothing is between the mover and moved with respect to place.

But surely this is obvious even from their definitions; for pushing is motion away from oneself or something else, toward some other thing, and pulling is motion away from some *10* other thing, toward oneself or something else, whenever the motion of the thing pulling is faster than the separating from another of the continuous things, for in this way the one is pulled along with the other. (But perhaps it might seem that there is also some pulling in another way, for it is not in this way that wood pulls fire. But it makes no difference whether the thing pulling pulls while moving or staying still, since when it does the one it pulls where it is, and when it does the other it pulls where it was.) But it is impossible to move anything either from oneself to something else or from something else to oneself *244b* without touching it, and so it is clear that nothing is between the mover and the moved with respect to place.

But surely neither is anything between what is altered and what alters it. And this is evident from considering examples; for in every one it turns out that the end of the thing causing alteration and the beginning of the thing being altered are together. For it is our assumption that it is something affected in the so-called passive attributes that is being altered. For in every case, body differs from body in perceptible attributes, with either more or fewer of them, or more or less of each of them; but surely also what is altered is altered by way of the things mentioned, since these are attributes of an underlying quality. For we say that something heated or sweetened or thickened or dried or whitened is altered, speaking in the same way of both the soulless and the ensouled, and of the ensouled things *10* in turn, of both the non-perceptive parts and the senses themselves. For in a certain way, the senses too are altered, since sensing, when it is active, is a motion through the body, with the sense being acted on in some way. In as many ways as the soulless thing is altered, then, *245a* the ensouled is as well, but the soulless thing is not altered in all the ways in which the ensouled thing is (since it is not altered by way of senses); and the one is unaware, the other not unaware, of being affected, though nothing prevents the ensouled thing too from being unaware, when the alteration happens not by way of the senses. Then so long as the thing altered is altered by means of perceptible things, in all of these cases it is evident that the end of the thing causing alteration and the beginning of the thing being altered are together; for with the one, the air is continuous, and with the air, the body. Again, color is continuous with light, and the light with the eyesight. And hearing and the sense of smell work in the *10* same way, for the first mover next to the thing moved is the air. And similarly with taste: for the juice is together with sense of taste. And it is the same way as well with soulless and non-perceptive things. Therefore nothing is between the thing altered and the thing causing alteration.

Nor surely is there anything between what is increased and what increases it; for the primary thing that causes increase causes it by coming to be present, so that the whole becomes one. And in turn, what causes decrease causes it by the becoming-absent of something belonging to the thing that is decreased. Then both what causes increase and what causes decrease must be *245b* continuous things, and nothing is between what is continuous. It is clear then that nothing is in the middle between the thing moved and the last mover, the first one from the thing moved.

Chapter 3

That every altered thing is altered by perceptible things, and that alteration is present only in those things which are said in their own right to be acted on by perceptible things, must be examined on the basis of the following things. One might suppose that alteration is present more than in other things among shapes or forms, and among active states, both in the taking on and losing of these; but it is present in neither of these. For *10* when the thing that is shaped and arranged is completed, we do not say that it is what it is made of, for example, that the statue is bronze, or the taper wax, or the bed wood, but using derivative terms we call them brazen, waxen, or wooden. But we do call a thing that has been affected and altered by the name of its attribute, for we say that the bronze or wax is fluid or hot or hard (and not only that way, but we also call the fluid or hot thing bronze), calling the material by the *246a* same name as the attribute. If, therefore, we do not call a thing that has come into being, in which a shape is, after its shape or form, but we do call things after their attributes and alterations, it is clear that comings-into-being could not be alterations. And even to speak in this way would seem to be absurd, to say that the human being or the house or anything else that has come into being has been altered; in order that each thing come into being, it is perhaps necessary that *something* be altered, such as the material that is dense or rare, or hot or cold, but it is surely not the things that come into being that are altered, nor is the coming-into-being of them an alteration.

10 But surely neither are active states alterations, neither those of the body nor those of the soul. For some of the active states are virtues and others vices, but it is not possible that either a virtue or a vice be an alteration; rather, the virtue is a certain perfection (for each thing is said to be complete when it takes on its excellence—for it is then most in accord with its nature—just as a circle is perfect when it has most of all become a circle and when it is best), and a vice is a spoiling or loss of this. Then just as neither do we call the completion of a house an alteration (for it would be absurd if the top course of stones or *20* the roof tiles were alterations, or if in being built to the top or tiled, the house were altered *246b* but not completed), it is the same way also with virtues and vices, and with the things that have them or take them on, for the one kind are perfections and the other losses, and so are not alterations.

Further, we say that all excellences consist in holding a certain relation. For those of the body, such as health and fitness, we place in the blending and due measure of the hot and the cold, either of themselves in relation to themselves in the things within, or in relation to their surroundings; and similarly with beauty and strength and the other excellences and *10* defects. For each consists in holding a certain relation, and disposes the thing having it well or badly toward its proper attributes, proper being those by which it by nature comes into being or is destroyed. Since, then, relations are not themselves alterations, nor is there alteration of them, nor becoming, nor in general any change at all, it is clear that neither active states nor the losing or taking on of active states are alterations, though in order that they come into being or be destroyed it is perhaps necessary that some things be altered, exactly as with the form or shape, such as the hot and the cold, or the dry and the moist, or those things in which these happen first to be present. For each defect or excellence is spoken of in relation to those things by which the thing having it is of such a nature as to be altered; for the excellence makes it be either unaffected or subject to be affected in just *20* a certain way, while the defect makes it contrarily subject to be affected or unaffected.

247a And it is the same with the active states of the soul, since these all consist in holding certain relations, and the virtues are perfections, the vices losses. And further, the virtue disposes something well toward its proper attributes, and the vice disposes it badly. Therefore, virtues and vices will not be alterations either, nor will the losses or takings on of them. But in order that they come into being, it is necessary that the perceptive part be altered. And this will have been altered by perceptible things, for all moral virtue is involved with bodily pleasures and pains, and these are present in acting, or in remembering, or in expecting. *10* Some, then, are in the action, following upon the sense perception, so that they are set in motion by some perceptible thing, while others are in the memory or the expectation derived from this, since they are pleasing when one is remembering what sort of things were experienced or anticipating what sort are going to be. So every such pleasure must come into being by way of perceptible things. And since when pleasure and pain come to be present, vice and virtue come to be present as well (for they are involved with these), while the pleasures and pains are alterations of the perceptive part, it is clear that something must be altered both for these to be cast off and for them to be taken on. Therefore, the coming into being of them follows an alteration, but they are not alterations.

247b But surely neither are the active states of the thinking part alterations, nor is there a coming into being of them. For most of all by far do we say that what has knowledge does so by holding a certain relation. And further, it is evident that there is no coming into being of these states, for what is potentially knowing becomes knowing not by being itself moved in any way, but by the becoming-present of something else. For whenever a particular thing has happened, the thinking part of the soul knows the universals in a certain way through the particular. And again, of the use and being-at-work of knowledge there is no coming into being, unless one thinks there is coming *10* into being of seeing and touching, for the using and being-at-work of it is similar to these. And the taking on of knowledge in the first place is not a coming-into-being or an alteration; for it is by the coming to rest and standing still of the thinking part that we are said to know and understand, and there is no coming-into-being into being at rest, for as a whole, there is none of any change, just as was said before. And further, just as when someone is set free from being drunk or from sleeping or from being sick, into their opposites, we do not say that someone has become knowing again (even though he was incapable of using his knowledge before), nor in the same way when in the first place one takes on the active state; for it is by the soul's calming down out of its native disorder that it becomes something *248a* understanding and knowing. For this reason too, children are able neither to learn nor to judge from sense perceptions in the same way as their elders, for their disorder and motion are great. The soul is calmed and brought to rest for some by nature itself, for others by other people, but in both kinds by the being altered of something in the body, just as in the case of the use and being-at-work, when one has become sober or has been awakened. It is clear, then, from the things said that being altered and alteration come about in perceptible things and in the perceptive part of the soul, but in no other thing, except incidentally.

Chapter 4

10 One might be at an impasse whether every motion is comparable with every other one or not. Now if they are all comparable, and if something of the same speed is moved an equal amount in an equal time, there will be something circular equal to a straight line, as well as greater or less. What is more, an alteration and some change of place would be equal, whenever in an equal time one thing is altered and another carried along. Therefore there would be a condition equal to a length, but this is impossible. But is there not an equal speed just whenever a thing is moved an equal *amount* in an equal time, while there is no condition equal to a length, and thus no alteration equal to or less than a change of place, so that not all motions are comparable?

But how will the answer turn out for the circle and the straight line? For it would *20* be absurd if it were not possible for this thing to be moved in a circle at the same speed as that one in a straight line, but it must automatically be moved either faster or more slowly, as if one were downhill and the other uphill. Nor would it make any difference to the argument if someone said that one must automatically be moved either faster or more slowly, for then there will be a circular line greater or less than a straight line, and therefore *248b* also equal. For if in the time A, one moves through B and the other through C, B would be greater than C, since that is what faster meant. Accordingly, too, if it is moved an equal amount in a lesser time, it is faster; and so there will be some part of A in which the distance B of the circle is equal to the C gone through in the whole of A. But surely if the motions are comparable, what was just said follows, that a straight line is equal to a circle. But the lines are not comparable; therefore neither are the motions, but things with a common name that does not have the same meaning are all incomparable. For example, why is there not a comparison as to which is sharper, a stylus, wine, or a high note? *10* It is because they are ambiguous, not comparable; but the high note is comparable with a neighboring tone, because sharp means the same thing for them both. Then speed is not the same thing in this case and in that, is it? And it is even less so in turn in alteration and change of place.

Or, first of all, is this not true, that if things are given a common name non-ambiguously, they are comparable? For "a lot of" means the same thing for water and air, yet they are not comparable. Or if "a lot of" is not unambiguous, "double" at any rate means the same thing (for it is two in relation to one), and all doubles are not comparable. Or is it also the same argument with these? And "a lot of" *is* ambiguous. But for some-things even the definitions are ambiguous, as, if one should say that "a lot of" is so much or more, the so-much differs; and "equal" is ambiguous, and, if it so happens, even "one" is ambiguous, *20* as an immediate consequence. And if this is, so is "two," since why would some things be comparable and others not if there were one nature?

Or is it because they are in different primary recipients? So a horse and a dog are comparable as to which is whiter (for that in which this primarily is, the outer surface, is the same), and similarly as to size; but water and a voice are not comparable as to which is clearer, since the thing compared is *249a* in different recipients. Or is it not obvious that it would be possible in this way to make everything one, but to say that each instance of it is just in a different recipient, and equal and sweet and white would be the same, but just in different recipients? Yet it is not any random thing that is receptive, but one thing primarily of one attribute.

But then must things that are comparable not only not be ambiguous, but not even have a difference in either what is compared or that in which it is? I mean, for example, that color has a division, and so things are not comparable in respect to this (such as which is more colored, not in respect to some particular color but just as color) but in respect to whiteness. So too with motion, what has the same speed is moved in an equal time a certain *10* equal amount; so if in this time some of the length is altered and the other part is carried along, is this alteration equal to and of the same speed as the change of place? It is absurd. And the reason is that motion has forms; so then if things carried through an equal length in an equal time will be of equal speed, the straight line and the circular one will be equal. Which is the reason, then: that change of place is a kind, or that a line is a kind? For the time is the same, but if the lines are different in form, those changes of place too differ in form. For change of place also has forms if that along which it is moved has forms. (And if there are times when that by-means-of-which differs, as, if feet, walking, but if wings, flying; or are these not distinct forms, but is the change of place different by means of the shapes?) *20* Therefore things moved through the same magnitude in an equal time are of equal speed, but the same magnitude has to be non-differing in form as well as non-differing in motion. So this must be examined: what a difference of motion is. And this argument implies that a kind is not some one thing, but a manyness is within it unnoticed, and some of the ambiguities that there are hold a great distance from each other, while others have some similarity, and yet others are very near either in kind or by analogy, on which account they do not seem to be ambiguous. When, then, is the form different: when the same thing is in something different, or when a different thing is in something different? And what is the dividing line? And by what do we judge that the white or the sweet is the same or different in form—is it because it manifests itself differently in something different, or because it is not the same at all?

30 And about alteration in particular, how will one be of the same speed as another? If being healed is being altered, one person is cured quickly, another slowly, and some *249b* people in the same time, so that an alteration of equal speed will be possible, since something is altered in an equal time. But what is altered? For the equal is not applicable there, but what equality is in the how-much, being alike is there. But let it be the case that what makes the same change in an equal time is of equal speed. Which, then, must the sameness go along with: that in which the attribute is or the attribute? In this case, it is because health is the same that it is possible to grasp that neither more or less but the same change is present. But if the attributes were different, as when a thing whitened and a thing healed *10* are altered, there is in these nothing the same or equal or alike, for which reason it immediately makes these into forms of alteration, and alteration is not *one*, just as neither are changes of place. So one must find out how many forms there are of alteration, and how many of change of place. So if the things moved differ in species, and have their motions in their own right and not incidentally, the motions too will differ in species; if in genus, the motions too in genus, and if in number, in number. But if the alterations are to be of equal speed, must one look at the attribute, whether it is the same or alike, or at the thing altered, whether, say, this much of one has been whitened, but that much of the other? Or must one look at both, so that the alteration is the same or different by means of the attribute, as that is the same or not the same, but is equal or unequal according as the *20* thing altered is equal or unequal? And with coming into being and destruction, one must consider the same thing. In what way is a coming-into-being of equal speed? It is so if in an equal time what comes to be is the same and indivisible, such as a human being, not just an animal; but one is faster if what comes to be in an equal time is different (for we do not have any two words in which to name this difference as with unlikeness), or, if thinghood were a number, if what came to be were a greater and a lesser number of the same form. But there is no name for what is common, any more than for each of the two, in the way that the surpassing attribute, or the one of higher degree, is more, while with the how-much it is greater.

Chapter 5

Since a mover always moves something, in something, and to some extent (and by in something, I mean in time, and by to some extent, I mean that there is some how-much of *30* a distance, since always at the same time it is moving something, it also has moved it, so that there will be some so-much through which the thing has been moved, and in so much time), *250a* if A is the mover, and B the moved which has been moved as far as the length C in a time as much as D, then in an equal time, an equal power to that of A will move half of B through the double of C, or move it through C in half the time D, for in this way it will be proportional. And if the same power moves the same thing this far in this time, and half the thing that far in half the time, then also half the strength will move half the body an equal distance in an equal time. For example, let E be half the power of A and let F be half the body B; then they stand similarly, and the strength is proportional to the heaviness, so that they will move an equal distance in an equal time.

10 Yet if E moves F through C in time D, E will not necessarily move the double of F through half of C in an equal time; so if A moves B as far as C in time D, then E, which is half of A, would not move B, in time D or any part of D, through any part of C in the same ratio to the whole as E is to A: for it will perhaps not move it at all. For it is not the case, if the whole strength moved something so much, that the half will move it either any amount or in any time whatever; for one person could move a ship if the strength of the ship-haulers or the distance they all moved it were divided by the number of them. For this reason the *20* claim of Zeno is not true, that any part whatever of a kernel of milled grain makes a sound; for nothing prevents it from not moving in any amount of time the air which the whole bushel moved when it fell. Indeed, if it were by itself, it would not move so much a part of the air as it would move when belonging to the whole. For it is not anything at all except as potentially in a whole. But if there are two things, and each of them moves each of two things so far in so much time, then when the powers are put together, they will move the composite of the burdens an equal distance in an equal time, for there is a proportion here.

So then is it the same way also with alteration and increase? For there is something *30* that causes increase and something that is increased, and in so much time, the one causes, *250b* and the other undergoes, so much increase. And what causes alteration and what is altered are similar—something *may* be altered so much with respect to more or less, and in so much time, and if in double the time, twice as much, or if twice as much, in double the time, while if half as much, in half the time (or if in half the time, half as much), or in an equal time, twice as much. But if the thing that causes alteration should cause so much in so much time, or the thing that causes increase cause so much in so much time, it is not necessary that either cause half as much in half the time, or if it cause half as much that it be in half the time, but it may perhaps be that it will not cause alteration or increase at all, just as with moving the heavy thing.

Commentary on Book VII

After the high level of generality of the discussion in Book VI, which continues in the first chapter of this book, in the bulk of Book VII Aristotle turns his attention to the rich array of concrete examples of motion we meet in our experience. Change of place, alteration, and growth are not here merged into a common description of continuous motion, but are broken down into their own kinds. Motions A, B, and C give way to pushing, pulling, whirling, tasting the juice from food and then becoming one with it, putting the roof on a house, hauling a ship, becoming moderate about pleasures, and calming down from the soul's disorder into understanding and knowing. The conclusions reached are still about all motions in common, but the reasoning is not from common properties but from a review of kinds. The examples are vivid and are chosen to shed light beyond themselves, since, as Aristotle reminds us (247b, 5–6), the soul knows the universal in and through particulars.

Nature is a cause that works from within, but Book VII turns to external sources of motion. This does not necessarily lead away from nature, since it was noticed at the end of II, 2, that a human being is generated not only by parents but also by the sun. The form acts as a cause for the sake of which the being acts, but parts of the natural environment contribute to its life as external movers that work on its material. About such outer sources of motion, Aristotle makes three main points. For every motion that comes from outside the moving thing, there must be a first mover, no matter how long the chain of intermediate causes, and there must be a last mover, together with and touching the body of the thing moved. Second, the power of such a mover cannot in general be known from its effects, since these are comparable only among things of similar kinds, changing in the same respect, and, in the case of change of place, moving along comparable paths; there is no way to say whether the sun changed the color of one person's skin more or less, or faster or more slowly, than it cured someone else of an illness. Third, the strength of the mover and magnitude of its effect cannot in general be broken down into proportional parts that predict anything true about the world; if a hundred men straining at ropes can drag a ship a hundred yards across an isthmus, it doesn't follow that one of them by himself, straining just as hard, can pull the ship one yard, since he will not be able to move it at all.

The second and third points show that a mathematical science of motion would have to falsify the world of experience, since even where we can measure magnitudes of things and their attributes, those magnitudes do not enter into all the relations that mathematical magnitudes do. Galileo initiates such a science by declaring speed to be a magnitude subject to all the propositions of Euclid's book on ratios, but qualitative changes, though they manifestly occur at various speeds, cannot be brought into the science. Newton declares that any part of a force impressed on a body always produces a proportional part of the change of motion, but he must qualify this by saying that it never happens in the sensible world, where the contrary force of friction can blot out the effect altogether, but only if the body is isolated in a void. These two thinkers give the mathematical use of the imagination precedence over all other ways of knowing. Aristotle uses the imagination in an entirely different way, that can see universal conclusions through examples, and ground those conclusions in sensory experience.

The second and third points also suggest that the variety of kinds of motion prevents the whole of nature from being brought together into a single account of the relation of movers to moving things, but the first point suggests otherwise. In each kind of motion the external mover gains access to the movable thing by bodily contact. The earlier books have narrowed the most proper sense of motion from four to three to two of its kinds, and at the beginning of VII, 2, Aristotle says in passing and without explanation that change of place is somehow the first of the kinds of motion. The common necessity for an external mover to touch the thing it sets into any of the kinds of motion might converge with what appears to be a hierarchical order among the kinds of motions themselves. Book VII does not raise the question whether the external movers that act on any being reflect the unity of the one cosmos, but it brings the inquiry to the threshold of that topic, which is taken up in Book VIII.

Book VIII, Chapters 1–6

Deduction of Motionless First Mover

Chapter 1

250b, 11 Did motion at some time come into being, not having been present before, and is it in turn passing away, so that there will be no motion, or is it something that neither has come into being nor passes away, but always was and always will be, belonging to beings in a way that is without death or pause, as though it were a kind of life for all things put together by nature? *That* motion is, everyone says who says anything about nature, because what they are all considering is the forming of the cosmos and what concerns coming into being and destruction, which cannot be if there is no motion. And those who say there are *20* infinitely many worlds, some of them coming into being and others being destroyed, say there is always motion (since their comings into being and destructions must be with motion), while those who say it is one and either everlasting or not, assume also about motion what corresponds to the account of the cosmos. But if it is possible at any time that there not be motion, this must come to pass in one of two ways, either as Anaxagoras says (for he says that after all things were together and at rest for an infinite time, the intelligence introduced motion and made things separate), or as Empedocles does, that the world is in part moving and at rest in turns, moving whenever love makes one out of many or strife makes many out of one, but at rest in the meantime, saying,

30 So in that a one has learned to grow out of many,
And many in turn are achieved when the one grows apart,
251a In this way things are becoming and have no stable life;
But since they never leave off changing constantly,
In this way they are always motionless in a cycle.

For by "changing" one must assume he means from one motion to another. One must investigate how these things are, for they are preliminary to the main business, not only toward seeing the truth in the contemplation of nature, but also toward the pursuit of the first source of things.

And let us begin first with the things that have been marked out for us in the earlier *10* books on natural things. Now we say that motion is a being-at-work of the movable, as movable. It is necessary, therefore, that there be present things that have the potency to move in accordance with each motion. And even apart from the definition of motion, everyone agrees that there must be something capable of moving with each motion for there to be motion, as, for something to be altered, the alterable, and for something to be carried along, that which is changeable with respect to place, so that there must be something burnable before being burned and something that can set it on fire before setting it on fire. Then too, these things must either have come into being at some time when they had not been, or else be always. But if each of the movable things came into being, then prior to any that might be taken, another change or motion must have come about, by which the *20* thing capable of being moved or causing motion came into being; but if they were always present beforehand when there was no motion, this would seem unreasonable as soon as one understands it, even though it would necessarily follow still more once one has gone further. For if, when there were some movable things and others capable of causing motion, at one time there were some first mover and first thing moved, but at another time nothing but rest, these things must have changed beforehand; for there was some cause of the rest, since rest is the deprivation of motion. So then before the first change there would be a previous change. For some things cause motion in only one way, but others cause opposite motions, *30* as fire heats but does not cool, while knowledge seems to be one cause of opposite effects. And there seems to be something of the same sort even in the former case; for a cold thing in a certain way heats something when it turns around and goes away, just as one who knows willfully *251b* makes a mistake when he uses his knowledge in a way that goes against its grain. But those things that are capable of acting and being acted upon, or moving and being moved, are not capable of it under all circumstances, but when they are in a certain condition and come near each other. So whenever they do come near, the one moves and the other is moved, whenever also the fact is that the one was so disposed as to cause motion and the other was movable. If, then, they were not always moving, it is clear that they were not in such a condition that they were capable, the one of being moved, the other of causing motion, but one or the other of them had to change; for this follows necessarily among relative things, as, if what is not now double is to be double, there be a change, if not of both things, of one *10* or the other. Therefore there would be some change previous to the first one.

On top of these things, how would there be a before and an after when there was not time? Or time when there was not motion? But if time is a number of motion or a sort of motion, if time always is, it is necessary too that motion be everlasting. But surely about time, except for one person, everybody seems to be of one mind, for they say it is ungenerated. And by this means, Democritus even proves that it is possible for all things to have come into being, since time is ungenerated. Plato alone makes it come into being, *20* for he says it is inseparable from the heavens, and that the heavens came into being. Then if it is impossible both that time be and that it be known without the now, while the now is a certain kind of middle thing, and a thing which holds a beginning and an end together, the beginning of the time that is going to be and the end of what has gone past, it is necessary that there always be time. For the extremity of the last time that is taken will be in some now (since there is nothing in time *to* take aside from the now), and so, since the now is a beginning and an end, there must always be time on both sides of it. But surely if there is time, it is clear that there is necessarily also motion, if time is some attribute of motion.

The same argument also concerns the being-indestructible of motion, for in the very *30* same way that it followed from the coming into being of motion that there would be some change previous to the first one, in the same way on that assumption there would be one after the last. For a thing does not stop moving and stop being movable at the same time, *252a* such as being burned and being burnable (since a thing admits of being burnable when it is not being burned), nor does it stop being such as to cause motion and stop causing motion at the same time. So what caused something to be destroyed would have to be destroyed when it had destroyed the other thing, and so in turn afterward with what caused this to be destroyed, for destruction too is a kind of change. But if these things are impossible, it is clear that motion is everlasting, and is not something that was at one time but at another was not. In fact, to speak in that way is more suited to fiction.

So too is saying that things are a certain way by nature, and one must accept this as a starting point, as Empedocles seems to say, declaring that by necessity it belongs to things *10* for love and strife in turn to rule and cause motion, while they rest in the mean times. And probably also those who make one original being, such as Anaxagoras, would speak this way. But surely there is nothing irregular among things that are by nature or result from nature, for nature is for all things a cause of order. But the infinite in relation to the infinite has no ratio, but every ordering is a pattern. But to be at rest for an infinite time, and then to be moved at some time which is in no way different from any other, that it should be now rather than earlier, and not even to have any ordering besides, is no work of nature any more. For what is by nature is either simply disposed in a certain way, and is not this way at one time but different at another, such as fire is carried upward by nature, and not at one *20* time so but at another not; or what is not simple has a pattern. For this reason the better way is that of Empedocles, and of anyone else who has said things are this way, that everything is at rest and in motion in turns, for such a thing already has some order. But it is also necessary for the one saying this not just to say it, but also to state the cause of it, and not to lay anything down or deem anything worthy of belief without an account, but to offer either examples or demonstration; for the things assumed are not themselves causes [why everything is at rest and in motion in turns], nor is this what it is for love or strife to be, but it belongs to the one to bring together, to the other to separate. But if the "in turns" is also to be included within the boundary, one must say in what cases it is thus, as that there is something which brings people together, love, and that enemies flee from each other; for *30* this is assumed to be present also in the whole of things, since it is obviously so in some cases. And the "for equal times" needs some explanation. And in general, to accept as a sufficient starting point that something always either is or happens in a certain way, is not to take things up in the right way, though it is as far as Democritus leads back the causes *252b* involved in nature: that it also happened this way before. But he does not think it worthwhile to seek the source of this "always," speaking rightly for some things, but not rightly that it applies to everything. For a triangle always has its angles equal to two right angles, but still there is some other cause of this everlastingness; yet of starting points there is no other cause of their being everlasting.

Then to the effect that there neither was nor will be a time when motion either was not or will not be, let this much have been said.

Chapter 2

The contraries to these things are not difficult to refute. But it might seem *10* to those considering it that there was motion when there had been none at all, most of all from things such as these: First, that no change is everlasting, for every change is from something to something, so that there must be, as limits of every change, the contraries between which it takes place, while nothing is moved to infinity. Further, we see that it is possible for something to be moved that neither is moving nor has in itself any motion, such as with soulless things of which neither any part nor the whole is moving, but from being at rest they are at one time set in motion, though it would be proper to them to be in motion either always or never, if motion does not come into being when it has not been present. But such a thing is evident most by far with ensouled things; for sometimes when there is no motion at all present in us, though we are still we are nevertheless at one time set in *20* motion, and a beginning of motion comes to be present in us from out of ourselves, though nothing outside moved it. For we see nothing like this with soulless things, but something else always moves them from outside, but we say that the animal itself moves itself. So if at one time it is completely at rest, then in a motionless thing a motion would come into being out of itself and not from outside. But if it is possible for this to happen in an animal, what prevents the same thing from happening also with the whole of things? For if it happens in a small cosmos, it would also in a big one, and if in the cosmos, then also in the infinite, if the infinite admits of being moved or being at rest as a whole.

30 Now of these things, the first one mentioned, that motion toward opposites is not always the same and one in number, is rightly said. For this is perhaps a necessity, if the motion of what is one and the same is capable of not always being one and the same; I mean, for example, whether there is one and the same sound from one lyre-string, or always a different one, though the string is in the same condition and moved in the same way. But *253a* however this may be, still nothing prevents some motion from being the same by being continuous and everlasting; this will be more clear from what comes later. The being moved of what was not moving is in no way strange, if the external mover is present at one time, but at another not present. Nevertheless, how this might be so is something one must examine, I mean such that the same thing is at one time moved and at another not moved by the same mover; for the one saying this is raising as an impasse nothing other than this: why it is that some things are not always at rest and others always in motion. But the third thing would seem to be most an impasse, that motion comes to inhere in that within which it was not *10* present before, the thing that happens with ensouled things; for having been at rest beforehand, afterwards the ensouled thing walks, having been moved by nothing outside it, as it seems. But this is false. For we always see something moved in the animal, of the parts congenital to it; but the cause of the motion of this is not the animal itself, but perhaps its surroundings. We say it moves itself not in the case of every one of its motions, but in the case of those in respect to place. So nothing prevents, but it is perhaps rather a necessity, that many motions come to be present in the body by means of the surroundings, while some of these set in motion thinking or desire, and that presently sets in motion the whole animal, such as happens with *20* those that are asleep; for even though no motion of perceiving is present, because some motion is nevertheless present, the animals wake up again. But about these things too, it will be clearer from what follows.

Chapter 3

The starting point of the examination is the thing that was also involved in the impasse spoken of: why some things are at one time moved, at another time in turn at rest. Necessarily either everything is always at rest, or everything is always in motion, or some things are in motion and some at rest, and of these last in turn, either the ones in motion are always in motion and the ones at rest always at rest, or everything is alike of such a nature as to be moved and be at rest, or there is yet a third thing left. For it is possible that some beings *30* are always motionless, and others always in motion, while others have a share in both, which is the very thing we must say, for this both holds the resolution of all the impasses and is the end for us of this present business. To say that all things are at rest, and to seek a reason for this, leaving sense perception aside, would be a feeble kind of thinking, and a dispute *253b* involving a whole and not a part, disputing the claims not only of natural but of every knowledge, so to speak, and of every opinion, since they all make use of what is in motion. And further, just as in writing about mathematics, things that threaten his starting points are not in the mathematician's way, and similarly too with other things, so neither does the thing now spoken about stand in the way of the one who studies nature, since it is a presupposition that nature is a source of motion.

And to say that everything is in motion is almost as false, though less opposed to this *10* pursuit; for it was set down in the writings on nature that nature is a source of rest, just as of motion, though motion is equally natural. And some people assert that motion is not of some things and not others, but of all things and always, but say this escapes the notice of our senses; and it is not difficult to go out to engage with these people, even though they do not distinguish what sort of motion they mean, or whether they mean all sorts. For neither increase nor decrease can go on continuously, but there is also the mean. And there is the argument similar to this one about the wearing down of stones by dripping, or the breaking apart of them by what grows out of them; for it is not the case, if something has pushed out so far or the dripping has taken away so much, that half as much was [added or taken away] beforehand in half the time, but just as with the ship hauling [250a, 9–b, 7], so many drips move so much, but a part of them *20* cannot move so much in any amount of time. What is taken away is divisible into many parts, but none of them is moved separately, but all together. It is clear then that it is not necessary for something always to be going away just because a decrease is infinitely divisible, but it goes away at one time as a whole. And similarly too with alteration of any kind whatever: for if the thing altered is infinitely divisible, it is not the case that for this reason the alteration is too, but it often happens all at once, as freezing does. Also, whenever something gets sick, there must come a time in which it gets well, and it is not transformed in the limit of the time [during which it sickens], and must change into health and not into anything else. *30* So to say alteration goes on continuously is to disagree too much with the obvious. For alteration is into contraries; and a stone becomes neither harder nor softer. And as regards changing place, it would be a wonder if it escaped our notice whether the stone were being carried downward or staying still on the earth. Further, the earth and everything else stay in their *254a* proper places, and are forcibly moved from them; so if some of them are in their proper places, everything cannot be in motion with respect to place either.

That, then, it is impossible for everything to be either always in motion or always at rest, one might believe from these things and others like them. But surely neither is it possible that some things are always at rest, others always in motion, and nothing sometimes at rest and sometimes in motion. One must say that this is impossible in this way too, just as with the ways mentioned before (for we see the changes spoken of in the same instances), and that the one who disputes these things is fighting against the obvious; for there would *10* be no increase nor any forcible motion, if a thing were not set in motion contrary to nature from having been at rest beforehand. This argument also does away with coming into being and destruction. But it seems to everyone that being moved is just about the same as becoming something or being destroyed; for a thing becomes that, or comes to be in that, to which it changes, and passes away from being that, or being in that, from which it changes. It is clear, then, that some things are moved and some at rest at some times.

But one must now join up the opinion that everything is sometimes at rest, sometimes in motion, with the things said earlier. And one must make a beginning again from the things that have now been distinguished, in the same way that we began before. For either *20* everything is at rest, or everything in motion, or some beings at rest and some in motion. And if some are at rest and some in motion, necessarily either everything is sometimes at rest and sometimes in motion, or some always at rest and others in motion, while yet others are sometimes at rest and sometimes in motion. That, then, it is not possible for everything to be at rest, was said before, and we say it now too. For even if it is in truth in the way some say it is, that being is infinite and motionless, still it does not *appear* so to the senses, but many beings manifestly move. So if there is false seeming, or seeming at all, there is also motion, and also if there is imagination, or such a thing as seeming one way at one time *30* and another at another; for imagination and seeming seem to be motions of a sort. But to examine about this, and to inquire after a reason for things with which we are better off than to need a reason, is to judge badly of the better and the worse, of what is worthy of belief and what is not, and of what comes first and what does not. Similarly, it is impossible that everything is in motion, or some things always in motion and others always at rest. Against *254b* all these things, one belief is sufficient: for we see some things sometimes moving and sometimes at rest. So it is clear that it is alike impossible for everything to be at rest or to be in motion continuously, as for some things always to be in motion and others always at rest. What is left, then, is to consider whether everything is of such a sort as to be moved and be at rest, or some are of this sort, while some are always at rest and some always in motion; for this is what one must show us.

Chapter 4

Of the things that cause motion and are moved, some do so incidentally, others *10* in their own right: incidentally, for example, all those that do so by belonging to things that cause motion or are moved, and those that do so in consequence of a part, but in their own right, all those that do so not by belonging to something that moves or is moved, nor by means of the moving or being moved of any of their parts. And of those that do so in their own right, some do so by their own action, others by that of something else, and some do so by nature, others by constraint and contrary to nature. What is itself moved by itself is moved by nature, as is each of the animals. (For the animal is moved itself by itself, and whatever things have in themselves a source of motion, we say are moved by nature, and so the animal as a whole itself moves itself by nature, even though its body admits of being moved *20* both by and contrary to nature; it varies according to what sort of motion it happens to be moved with and what sort of elements it is made of.) But of things moved by something else, some are moved by nature and others contrary to nature: contrary to nature, for example, things made of earth are moved upward and fire downward, and moreover the parts of animals are often moved contrary to nature, on account of their positions or the manner in which they are moved. That a thing moved is moved by something is most clear in what is moved contrary to nature, since it is obviously moved by something else. Next clearest after motions contrary to nature are those of the things moved according to nature which are themselves moved by themselves, such as animals: for it is not whether it is moved by something that is unclear, but how one should distinguish within it the mover and the moved. *30* For it seems that just as in ships and things put together not by nature, so also in animals the mover is divided from the moved, and in this way the whole itself moves itself.

But the one that is left from the last distinctions mentioned is most an impasse; for of things moved by something else, we have set down some as being moved contrary to *255a* nature, while others remain to be set against these as by nature. And these are the ones that present the impasse as to whether they are moved by anything, for instance light and heavy things. For these are moved to the opposite places by force, but to their proper ones, the light up and the heavy down, by nature; but whether by something is no longer clear, as it is when they are moved contrary to nature. For to say that they themselves move themselves is impossible, for this is indicative of life and peculiar to things with souls, and they themselves would then be able to stop themselves (I mean, if it were responsible for its own walking, say, it would also be responsible for its not walking), so that if the being *10* carried upward of fire were in its own power, it is clear that it would also be in its power to be carried downward. And it is unreasonable that they would be moved by themselves with one motion only, if they themselves really do move themselves. What's more, how could something continuous and same-natured itself move itself? For insofar as it is one and continuous, not by contact, in this way it would be unaffected; but insofar as it is divided, in this way one part may be of such a nature as to act and the other to be acted upon. Therefore, none of these things itself moves itself either (since they are same-natured), nor does any other continuous thing, but there must be a division in each between the mover and the moved, as we *see* with soulless things, whenever any of the ensouled ones moves. But it turns out that these [light and heavy] things too are always moved by something, and *20* this would become clear to those who distinguish the causes. And the things spoken of are also graspable with the things that cause motion; for some of them tend to move things contrary to nature, as the lever, for example, is suited to move weights non-naturally, while others tend to move things by nature, as what is actively hot is suited to move what is potentially hot. And it is similar with the other distinctions of this kind. And movable by nature in just the same way are things that are potentially of-this-kind, this much, or here, whenever they have such a source in themselves, and not incidentally (for the same thing might be both of-this-kind and so-much, though the one is incidental to the other and does not belong to it in its own right). So fire and earth are moved by something by force when *30* they are moved contrary to nature, but are moved by nature whenever they are moved to be at work in the ways that belong to them potentially.

But since "potentially" is meant in more than one way, it is this that is responsible for its being unclear that such things are moved by something, fire upward, that is, and earth downward. Now the one who is learning is potentially knowing in a different way from the one who already has knowledge but is not at work with it. But always, whenever what can *255b* act and what can be acted upon are together, what is potential comes to be at work, as the one learning, from being something potentially, becomes something potentially in a different way (for the one having knowledge but not contemplating is potentially knowing in a certain way, but not in the way he was before having learned), and once he is in this condition, if nothing prevents it, he is at work and contemplates, or else he would be in the contradictory condition, that of ignorance. And these things are similar with natural things; for what is cold is potentially hot, but when it has changed, by the time it is fire, it burns, unless something prevents it and gets in the way. And the heavy and the light are the same way *10* too, for what is light comes into being from what is heavy, as air does from water (for it was this potentially first), and by the time it is light, it will immediately be at work, unless something prevents it. But the being-at-work of what is light is to be somewhere above, and it is being prevented whenever it is in the opposite place. And it is this same way too with what is so-much and what is of-this-kind.

However, it is this that is being sought, *why* the light and heavy things are moved to their own places. But the cause is that they are innately directed somewhere, and being light or heavy is just this, the former distinguished by the upward, the latter by the downward, direction. But a thing is potentially light or heavy in more than one sense, as was said; for when it is water it is potentially light in one way, and when it is air it is still in a way *20* potentially light (since it is possible, when it is impeded, that it not be up above). But if the impediment is taken away, it is at work and always becoming higher. And in a similar way the of-this-kind changes to being at-work, for the knower immediately contemplates unless something prevents it, and the so-much stretches out unless something prevents it. And there is a sense in which the one who moves a support or obstacle causes the thing to move, and another in which he does not, as in the case of one who pulls away a pillar or takes a stone off a wine-skin in the water; for he moves the thing incidentally, just as also the ball whose course is abruptly broken off and turned back is moved not by the wall but by the one *30* who threw it. It is clear, then, that none of these things itself moves itself, but it has a source of motion, not of causing or producing motion but of undergoing it. So if everything that is moved is moved either by nature or by constraint and contrary to nature, and everything that is moved by constraint and contrary to nature is moved by something other than itself, and of the things moved by nature in turn, both those moved by themselves and *256a* those not moved by themselves are moved by something, as are light and heavy things (for they are moved either by what brought them into being and made them light or heavy, or by what unfastened their impediments and obstacles), then everything that moves would be moved by something.

Chapter 5

But this happens in two ways; for it is either not on account of the mover itself, but on account of something else that moves the mover, or on account of it, and this either as first from the last thing or through a number of things, as the stick moves the stone and is moved by the hand, which is moved by the human being, while this no longer moves by being moved by something else. Now we say that both the last and the first of the *10* things moving something cause its motion, but the first more so, since it moves the last while the last does not move the first, and without the first the last would not move anything, while the first would without the last, as the stick would not move anything if the human being did not move it. But if everything that moves must be moved by something, and by something that is either moved by something else or not, and if it is moved by something else there is necessarily some first mover which is not moved by anything else (since it is impossible that the thing moving and itself being moved should go on to infinity, for of infinitely many things *20* none is first)—if, then, everything that moves is moved by something, while the first mover is indeed moved, but not by anything else, it must itself be moved by itself.

And it is possible for the same argument to come forward again in this way. For everything that causes motion both moves something and moves it with something. For the mover moves either with itself or with something else, for example, a human being either himself or with a stick, and the wind either knocks something down itself, or the stone which it pushes does. But it is impossible to move something unless the mover itself, with itself, moves that with which it causes the motion; but if it causes the motion with itself, there need not be anything else by which it causes it, and if there is a different thing with which it causes the motion, there is something which causes motion with nothing but itself, or else it would go to infinity. If, then, anything that is moved causes motion, it is necessary to stop *30* and not go to infinity; for if the stick causes motion by being moved by the hand, the hand moves the stick, and if something else causes the motion with this, something different from this as well is the mover. So whenever a different thing always causes a motion with something, there must be, taking precedence over them, a thing that causes the motion itself, with itself. And if this is moved, but no other thing is moving it, it must move itself; so also *256b* following this argument, a thing moved is either moved directly by something that moves itself, or comes at some point to such a thing.

And besides what has been said, these same things will also follow for those who examine in this way. For if everything that moves is moved by something that is moved, then this belongs to the things either incidentally, so that they cause motion while being moved but not on account of the being-moved itself, or not incidentally but in their own right. First, then, if it is incidental, it is not necessary that the thing causing motion be moved. But if this *10* is so, it is clear that it is possible at some time for none of the beings to be in motion, for what is incidental is not necessary but admits of not being so. But if we assume something possible to be the case, nothing impossible will follow, though perhaps something false will. But that motion not be is impossible, for it has been shown earlier that motion must always be. And this follows reasonably, for there must be three things, the moved, the mover, and that by which it causes motion. Now the moved must be moved, but need not cause motion; that with which it is moved both causes motion and is moved (since this changes along with and in the same way as the thing moved, being together with it; and this is evident with *20* things moved in respect to place, for they must touch each other to some extent); but that which causes motion in such a way that it is not that with which it causes it, is motionless. And since we see the one extreme, which can be moved but has no source of motion, as well as that which is moved not by anything else but by itself, it is reasonable—we do not say necessary—that there be also the third kind which causes motion while being motionless. And for this reason Anaxagoras speaks rightly when he says that intelligence is unaffected and unmixed, seeing that he makes it be the source of motion; for only this kind of source could cause motion while being motionless, and rule while not being in contact.

But now if that which causes motion is moved not incidentally but necessarily, and *30* could not cause motion if it were not moved, then the mover, to the extent it is moved, must be moved either with the same kind of motion that it causes, or with a different kind. I mean that either the thing heating is also itself heated and the thing healing is healed and the thing carrying carried, or else the thing healing something is carried and the thing carrying *257a* something is increased. But it is clear that this is impossible; for it is necessary to articulate a division all the way down to the indivisible kinds, as, if something teaches one to do geometry, that this itself be taught to do geometry, or if it throws something, that it be thrown with the same kind of throw. Or if it is not this way, then one motion must come from the other kind; for example, the thing carrying something is increased, but the thing that increases it is altered by something else, and the thing that alters that is moved with some different motion. But it is necessary to stop, since the kinds of motion are limited. But to turn it back on itself, and say that the thing causing alteration is carried, is to do the *10* same thing as if one said immediately that the thing carrying was being carried and the thing teaching was being taught (for it is clear that everything that moves is also moved by the mover higher up, and more so by the movers with more precedence). But surely this is impossible, for it implies that the one teaching is learning, of which activities, the one requires not having, the other having, knowledge.

And still more unreasonable than these things, it follows that everything that can move something is movable, if in fact everything that is moved is moved by something that is moved; for it would be movable, in just the same way as if one should say that everything that can heal something is healable and everything that can build something can be built, *20* either directly or through a number of steps. I mean, for example, if everything that can cause motion is movable by something else, but not movable with the same kind of motion with which it moves the next thing, but with a different kind, say, what can cause healing is capable of learning, still this, in going up to what is above it, comes at some point to the same kind, as we said before. So the one alternative is impossible and the other sounds like fiction; for it is absurd that what can cause alteration is necessarily such as to be increased. Therefore it is not necessary for what is moved always to be moved by something else, and for this to be moved; and therefore it will come to a stop. And so the first thing moved will be moved by something at rest, or it will move itself.

But surely if one should need to consider whether the cause and source of motion is *30* something which itself moves itself or something moved by another, everyone would vote for the former; for that which is itself derived from itself always takes precedence in responsibility over that which depends on both something else and itself. So this must be examined by those who take up another starting point: if something itself moves itself, in what respect and in what manner it does so.

Now everything that moves must be divisible into parts that are always divisible; for *257b* this was shown before in the general sections about nature, that everything moved in its own right is continuous [234b, 10–20]. And it is impossible for anything that itself moves itself to move itself as a whole; for the whole would be carried along and yet would be causing the same change of place, while being one and indivisible in kind, or would be altering and causing alteration, so that one would be teaching and learning at the same time, or producing healing and being healed to the same health. Further, it has been set down that it is what is movable that is moved; but this is something that is moved potentially, not actively-and-completely, and what is the case potentially goes over into being-active-and-complete, and motion is an incomplete being-active-and-complete° of the moveable thing. But the thing that causes motion is *10* already in activity, as what causes heat is what is hot, and in general what brings something into being is what has the form. So at the same time, the same thing, in the same respect, would be hot and not hot. And it is similar with every other one of those things of which the mover must have the same name. Therefore, of the thing that itself moves itself, one part causes the motion and another part is moved.

That it is not possible for a thing itself to move itself in such a way that each part is moved by the other, is clear from the following. For there will not be any first mover, if it does move itself in both ways (for what takes precedence is more responsible for the being moved, and is the mover to a greater degree, than the thing next after it; for there are two ways to move something, in one of which the mover is itself moved by something else, and *20* in the other by itself, while the one further from the thing moved is nearer to the source than is the one in between). Also, it is not necessary that the thing causing motion be moved except by itself; therefore it is incidental that the other part should cause motion in return. So I take the not moving to be possible: therefore there is one part that is moved and another that causes motion but is motionless. It is still not necessary that the thing causing motion be moved in return, but either something motionless must cause it, or it must be moved by itself, if there must always be a motion. Again, it would be moved with that motion which it causes, so that the thing causing heat would be heated.

But surely of the thing itself that primarily moves itself, neither one nor more than *30* one part could each itself move itself. For if the whole is itself moved by itself, it will be moved either by some one of its parts or as a whole by the whole. Then if it is moved by itself by way of the being-moved of some part, this part would be that which itself primarily moves itself (for when separated, this would itself move itself, but the whole no longer would); but if the whole is moved by the whole, these parts would themselves move *258a* themselves incidentally. So since this is not necessary, let the parts' not being moved by themselves be taken. Therefore, of the whole, one part will cause motion while it is motionless, and the other will be moved; for only in this way is it possible for something to be a self-mover. Again, if the whole itself moves itself, one part of it will cause the motion and the other will be moved. Therefore AB will be moved both by itself and by A.

And since one thing causes motion while being moved by something else, another while being motionless, and one thing is moved while moving something, another while moving nothing, the thing that itself moves itself must be made of something motionless but causing motion and also of something moved but not necessarily causing motion, but *10* whichever way it turns out. For let there be A, which causes motion but is motionless, B, which is moved by A and moves C, which is moved by B but does not move anything; for even if through more than one thing it comes at some point to C, let it be through one only. Now the whole ABC itself moves itself. But if I take away C, AB will itself move itself, A causing the motion and B being moved, but C will not itself move itself, nor will it be moved at all. But surely neither will BC itself move itself without A; for B causes motion by means of being moved by something else, not by way of any part of itself. Therefore only AB itself moves itself. Therefore, that which itself moves itself must have something that causes *20* motion but is motionless, and something that is moved but does not necessarily move anything, with either both touching each other or one of them touching the other.° Then if what causes motion is continuous (for what is moved must be continuous), each will touch the other. So it is clear that the whole itself moves itself not by means of some part's being such as itself to move itself, but as a whole it moves itself, being moved and moving by means of some part's causing motion and some part's being moved. For it neither causes motion nor is moved as a whole, but A causes the motion, and B is moved only.

But there is an impasse, if someone takes away part of A, if what causes motion but *30* is motionless is continuous, or part of the thing moved, B. Will the remainder of A cause motion and the remainder of B be moved? For if this is so, it would not be AB that is *258b* primarily moved by itself, since when something is taken away from AB, the remnant of AB still will move itself. But nothing prevents each part, or one of them, the one that is moved, from being divisible potentially but actually undivided; and if it were divided it would no longer have the same nature, so that nothing prevents it from being present primarily in things that are potentially divisible. It is clear, then, from these things that the first mover is motionless; for whether the thing that is moved, and moved by something, is situated immediately at the motionless first thing, or at something that is moved but itself causing itself to move and to stop, in both ways it turns out that for everything that moves, the thing that first causes the motion is motionless.

Chapter 6

10 And since there must always be motion and not be any gaps, there must be something everlasting that originates motion, whether it is one thing or more than one; and the first mover is motionless. Now for each thing that is both motionless and a cause of motion to be everlasting is in no way related to the present argument; but that there must be something motionless that is free from *all* change, both simple and incidental, yet is such as to cause the motion of something else, is clear to those who examine it in the following way. Now let it be possible with some things, if anyone so desires, to be sometimes and not to be sometimes, without coming into being or being destroyed (for it is perhaps necessary, if anything *20* without parts *is* at one time but at another time is not, for every such thing to be at one time and not be at another time without changing). And let it also be admitted as possible for some of the original motionless sources of motion to be at one time and not be at another time. But this is not something possible for all of them; for it is clear that there is something responsible for the being at one time and not being at another of the things that themselves move themselves. For everything that itself moves itself must have magnitude, if nothing without parts is moved, though there is no such necessity at all from what has been said for the thing that causes motion. For none of the things that are motionless, but are not always in being, is responsible for some things' coming into being and others' being destroyed and for this being so continuously, not even if some were responsible for them and others in turn *30* for *them*. For neither any of them nor all of them is responsible for what is everlasting and continuous; for to be so is to be everlasting and necessary, while they are infinite in number *259a* and not all present at the same time. So it is clear that even if some of the motionless causes of motion and many of the things that themselves move themselves are destroyed tens of thousands of times, while others come along in their place, and this motionless thing causes this thing to move but another one causes that one to, nevertheless there is still something that overarches and is beyond each of them, which is responsible for the being of some things and the not-being of others, and for continuous change; and this is responsible for them, and they for the motion of other things.

So if motion is everlasting, the first mover too will be everlasting, if it is one; if there are more than one, there will be more than one everlasting thing. But one ought to regard *10* it as one rather than many, or finite rather than infinite. For when the same results follow, one must always choose the finite in preference; for in things that are by nature, the finite and the better are bound to be present, if they are possible. But even one is sufficient, which, being the first among the everlasting, motionless things, would be the source of motion for everything else.

But it is clear also from the following that the first mover must be something one and everlasting. For it has been shown that there must always be motion. And if it always is, it must be continuous; for what is always so is continuous, while what is in a series is not continuous. But surely if it is continuous, it is one. But what is one is what is brought about by one mover and belongs to one moved thing; for if one thing after another were to cause *20* the motion, the whole motion would not be continuous but a series.

So one might be persuaded that there is a first motionless thing by the preceding things, and again by looking at those among the movers that originate motion. Now it is clear that there are some beings that are sometimes in motion and sometimes at rest. And by this means it became clear that neither is everything in motion, nor everything at rest, nor some things always at rest while the others are always in motion; for the things that do both and have the capacity to be moved or be at rest give the proof on the subject. But since things of this sort are obvious to everyone, we wish to show also for the two kinds the nature of each, that there are some things that are always motionless and others that are always in *30* motion. So passing on to this, having established that everything that moves is moved by something, that this is either motionless or in motion, and that when it is in motion it is always moved either by itself or by something else, we got as far as to grasp that among *259b* things in motion there is a source of their being moved which itself moves itself, and among all things a source that is motionless; and as for things which themselves move themselves, we see things that are obviously of this kind, namely the class of ensouled things and of animals. And they furnished us with the opinion that it was possible for motion to become present at one time when it had not been present at all, because of our seeing this happening in them (for, being motionless, they are moved again, as it seems), but it is necessary to grasp that they set themselves in motion by way of one kind of motion [253a, 14–20], and not as the origin even of this; for the cause is not from themselves, but other natural motions are present in the animals, which are not set in motion by them themselves, such as growing, *10* wasting away, and breathing, with which each of the animals is moved when it is at rest and not being moved with any motion by its own agency. What is responsible for this is the surroundings, and many of the things that enter inside, such as the food in some cases; for they sleep while it is being digested, and wake up and move themselves when it is dissolved into its elements, the first source of motion being external, for which reason they are not always moved by themselves continuously, since something else is the mover, itself being moved and changing in relation to each of the things that move themselves. And in every one of them, the first mover and cause of its moving itself by itself, is itself moved, though *20* incidentally; for the body changes its place, and so what is in the body also moves itself as though by a lever.°

From these things it is possible to be persuaded that, if something is one of the motionless things, but one that also moves itself incidentally, it is impossible for it to cause continuous motion. So if there must be motion continuously, there needs to be something that first causes the motion that is without motion of even an incidental kind, if, as we assert, there is going to be among beings some ceaseless and deathless motion, and if being itself is to remain self-contained and in the same condition; for since the source of motion stays as it is, the whole of things must also stay as it is, since it is continuous with the source. And for something to be moved incidentally by itself, and by something else, are not the same; *30* for to be so moved by something else belongs also to some of the sources of motion of things in the heavens, as many as are carried about in more than one change of place,° but to be so moved only in the other way belongs to destructible things.

But surely if there is something that is always of this kind, a mover that is itself *260a* motionless and everlasting, the first thing moved by this must also be everlasting. But this is clear also from there being no other way for there to be coming into being and destruction and change of the other things, if it were not the case that something that is moved moves them; for what is motionless always causes motion in the same way and with one motion, inasmuch as it in no way changes in relation to the thing moved. But something that is moved by something that is moved, even if that is moved directly by what is motionless, will, on account of its varying, be related to things in varying ways, and will not be responsible for a motion that is the same, but on account of being in opposite places or forms, it will *10* present itself in opposite ways to each of the other moved things, sometimes being at rest, and other times being in motion. But what has been said has also cleared up something that we were at an impasse about at the beginning; why it is that everything is not in motion or at rest, nor some things always in motion and the others always at rest, but some things are sometimes one way and sometimes not. For the cause of this is now clear, because some things are moved by what is motionless and everlasting, and hence are always moved, and others by what is moving and changing, so that they also must change. But what is motionless, as was said, inasmuch as it simply and in the same way persists in the same condition, will cause a motion that is one and simple.

Commentary on Book VIII, Chapters 1–6

In Plato's *Timaeus* we are told that the world, motion, and time all came into being. One cannot conclude that this is Plato's own opinion, since he puts it in the mouth of a character, and not even in that of Socrates, and also because Aristotle has told us before (209b,

12–13) that what Plato says in the *Timaeus* about place is not what he believed and taught. Timaeus himself says in the dialogue that before there was any time, there was a divine craftsman who looked at the eternal things and crafted copies of them, and Timaeus apologizes for the necessity of speaking in ways that are imprecise and self-contradictory (29 A–D). Aristotle shows that a first beginning of time, *before* which there was no time, and a first beginning of motion, into which movable things passed by a prior transition from immobility, are indeed self-contradictory thoughts, refuted as soon as they are formulated. Natural motion and change take place within constant patterns, and not in arbitrary fits and starts. In the temporal sense, there could not have been a first motion.

But there appear to be spontaneous transitions from rest to motion in nature, as when an animal wakes from sleep. Why couldn't the cosmos have had a first awakening? This analogy ignores all the motions going on in the sleeping animal—breathing, digesting, and so on—and the motions of its surroundings, which eventually contribute to waking it up. But those who say that everything is always in motion ignore equally plain facts: some changes are sudden, such as freezing, others come to an end, as getting sicker and sicker stops and gives way to getting well, and all movable things hold on in all sorts of constant conditions. Natural things display both motion and rest, but what governs the transitions from one state to the other? Even though a rock has an internal source of motion, it does not make sense to say it moves itself, since it does not control the start or stop of its motion. A falling rock is like a ball bouncing off a wall. The ball is not the source of its return flight, nor is the wall; the motion points back to the one who threw the ball in the first place. Similarly, the rock had to get up above its proper place before it could fall back. The being-at-work of the rock is less evident than that of fire, but it is constant, holding it down when it is on earth, straining downward when it is sitting on a table, and evident as a potency when it is falling. But the transitions between being down and at rest, being up and blocked by an obstacle, and being in downward flight all require external movers of some kind.

Where there is an external mover there is a first external mover. A baseball is hit by a bat, which is moved by hands, and we may insist on electrical impulses in the nerves and brain to move the hands, but no matter what analysis may be given of the chain of intermediate causes, the series must end exactly where Aristotle says it does, with the human being as a whole. This is not a series of events in time, since that has no beginning; the batter had to eat breakfast, before which he had to be born, before which his parents had to meet, and so on with no possible temporal beginning. The first mover is not an event but a being, and stands first not in time but in the order of responsibilty. Every such first mover must be motionless, or must move itself. The batter is in one sense motionless, since he stands in one place and pivots, but clearly something in him sets the parts of his body in motion. The true first mover is his soul, the being-at-work of his body as a whole, which is responsible for any motion that originates in him. Like the first mover in any sequence, that source is motionless not in the sense of being inert, but by being fully at-work, in an activity that is the same and complete at every instant. Those potencies in things which do not emerge by the inner activity of their own forms can still emerge by coming into contact with something already at-work.

From the standpoint of Aristotle's general understanding of being, motion can never be accounted for by other motions. Once physics is mathematized, motions can be explained *only* by other motions, or by forces or energies that are themselves defined in terms of motions, since the cause must equal the effect. In Aristotle's physics, a cause, to be a cause, must surpass the effect, and the relation is not quantitative. Every cause is a being at work that is fully what it is. A motion too is a being-at-work, but what it is fully is inherently not whole, but is a potency *of* some being. There was a mild formulation of this at 201b, 32–33, when motion was called an incomplete being at work (*atelēs energeia*), a phrase used often in Aristotle's works and explained as the complete being at work of something incomplete. Here (and nowhere else in his writings) Aristotle uses a deliberately self-contradictory formulation, calling motion an incomplete being complete (*atelēs entelēs-echein*), to emphasize that no motion can stand on its own as a source of anything, not even of itself. In the un-symmetrical relation between the being-at-work that causes motion, and the motion it causes, even the mutuality of the relation of touching can break down. The mover can touch the moving thing without being touched by it. In *On Generation and Corruption* (323a, 26–34), Aristotle gives the example of someone who causes us grief, but is unaffected by us, and in XII, 7, of the *Metaphysics* all sorts of familiar examples are suggested; the ice cream in the freezer, just being itself, or the beauty on the beach, unaware of our existence, can be a source of desire. Parmenides was wrong in making being annihilate motion, but being overarches, governs, and grounds all motion.

But if there are many things which can be motionless sources of motion, despite the fact that they come to be and pass away, and undergo all sorts of motions not related to the motions they cause, they themselves must point back to prior sources of their own motions and changes. The soul of a living thing is in one sense an ultimate source of its motions, but other, external sources must also contribute to its life; in an incidental way, the soul even undergoes change of place along with the body it moves, as though the body were a lever by which it moved itself. And the calm, orderly, repeating motions of the planets combine in themselves distinct primary motions, as Venus moves westward around the sky every day, eastward every year, and in a cycle from evening star to morning star and back again every seven months or so; each motion requires a distinct source. Is there a single, altogether motionless, first source of all motion in the world? If so, it could cause only a single, continuous, everlasting motion. Does anything move in that way? This question is taken up in the next and final section.

Book VIII, Chapters 7–10

The First Motion

Chapter 7

260a, 20 But now these things will be even more clear to those who make another start. For one must consider whether it is possible for any motion to be continuous or not, and if it is possible, what this motion is, and which of the motions is primary; for it is clear that, if there must always be motion, and a certain motion is primary and continuous, then the first mover moves something with this motion, which must be one and the same, continuous, and primary. But since there are three motions, with respect to size, attribute, or place, this last, which we call change of place, must be primary. For it is impossible for there to be *30* increase if alteration is not present beforehand; for what is increased is increased in one way by its like, but in another way by what is unlike, since the opposite is said to be food for its opposite. But everything becomes attached to its like by becoming like it. So there must *260a* be alteration for the change into an opposite. But surely if something is altered, there must be something that alters it, and makes it from, say, potentially hot into actively hot. So then it is clear that the thing causing the alteration does not hold on in the same condition, but is at one time nearer to, at another farther from, the thing that is altered. But these things could not be present without change of place. Therefore, if it is necessary that there always be motion, there must also always be change of place as the primary one of the motions, and of kinds of change of place, if one is primary and the others derivative, the primary one.

Besides, the origin of all attributes is the process of becoming dense or rare; for *10* heavy and light, soft and hard, and hot and cold seem to be certain kinds of thickness and thinness. And the process of becoming dense or rare is a combination or separation, by which the coming into being and destruction of independent things is said to take place. But things that are combined or separated must change place. But surely also the magnitude of what is increased or decreased changes in relation to place.

And besides, to those who look at it from the following direction, it will be clear that change of place is primary. For as with other things, so also with motion, *first* may be meant in more than one way. A thing is called more primary if other things will not be when it is *20* not, while it can be without the others, or if it is earlier in time, or in consequence of thinghood. So since there must be motion continuously, and there would be motion continuously if it is either continuous or one after another, but more so if it is continuous, and it is better for it to be continuous rather than one after another, and we always assume what is better to be present in nature so long as it is possible, and it is possible for it to be continuous (which will be shown later; let it be assumed now), and this can be no other motion than change of place, then necessarily change of place is primary. For there is no necessity at all for the thing that changes place to be either increased or altered, nor, certainly, to come into being or be destroyed; but none of these is possible if there is not the continuous process which the first mover sets in motion.

30 Besides, it is first in time; for everlasting things admit of being moved only in this motion. But with any one whatever of the things that have a coming-into-being, change of place must be the very last of the motions; for after they have come into being, there is first *261a* alteration and growth, while change of place is a motion of things that are already complete. But something else must earlier be moved with a change of place, and it is responsible for the coming into being of the things that come into being, without itself then coming into being, as what brings into being that which is brought into being, for otherwise it would seem that coming into being was the first of the motions for the reason that the thing must first have come into being. While it is that way with any one whatever of the things that come into being, still some other thing must be moved earlier than the things that come into being, which itself *is* and is not becoming, and another one before it. But since it is impossible that *10* coming into being be first (for then everything that moves would be destructible), it is clear that neither is any of the motions next in order earlier. By next in order, I mean increase, then alteration, decrease, and destruction; for they are all later than coming into being, so that if not even coming into being is earlier than change of place, neither is any of the other changes.

And in general what is becoming seems incomplete and headed toward its origin, so that what is later in the process of coming into being is more primary in nature. But change of place arises last for everything in the process of becoming. For this reason, some living things are completely motionless for lack of an organ, such as plants and many kinds of animals, while in others change of place arises as they are becoming complete. So if change of place belongs more to those things which have taken on their natures the more, this motion would *20* also be primary among the rest in relation to the thinghood of things, both for these reasons and because in changing place, least of the motions, does the moving thing depart from being what it is; for as a result of it alone there is no change of being, in the way that the of-what-sort of something altered, or the how-much of something increased or decreased, is changed. But most of all, it is clear that something that itself moves itself most authentically brings about this motion, the one that relates to place; but of things moved and causing motion, we say this one to be a source and first for things that move, namely the thing that itself moves itself.

That, then, change of place is primary among motions, is clear from these things; and *30* which change of place is the primary one must now be shown. And at the same time, the thing assumed both now and before, that it is possible for some motion to be continuous and everlasting, will be clear by the same line of thinking. Now, that none of the other motions admits of being continuous, is clear from the following. For all motions and changes are from opposites to opposites, as, with coming into being and destruction, the limits are being and not being, with alteration they are contrary attributes, and with increase and decrease, *261b* either greatness and smallness or completeness and incompleteness of size; and changes into opposites are opposite changes. But the thing that was previously in being, but is not always moved with the same motion, must previously have been at rest. It is clear, then, that in the opposite conditions the thing that changes will be at rest. And it is similar too with the changes; for destruction and coming-into-being are opposites simply, and each particular one is opposite to some particular one. So if it is impossible to be changing at the same time in opposite ways, the change will not be continuous, but a time will be between them. For it makes no difference whether changes to contradictories are contraries or not, so long as *10* it is impossible for them to be present to the same thing at the same time (for the former is not necessary to the argument), nor whether a thing need not be at rest in the contradictory, nor even whether change is contrary to rest (for what is not is perhaps not at rest, and destruction is into what is not), just so long as a time occurs between; for in this way the change is not continuous, and not in the earlier argument either was the being-contrary necessary, but rather the not being able to be present at the same time.

And one need not be alarmed that the same thing will be contrary to many things, for example, a motion both to standing still and to a motion to its contrary, but one need only grasp this, that the contrary motion is opposite in a certain way both to the motion and *20* to the state of rest, just as what is equal and the mean is opposite both to what exceeds it and what it exceeds, and that it is not possible for opposites, whether motions or changes, to be present at the same time. What's more, with coming into being and destruction, it would even seem to be completely absurd if what has come into being must immediately be destroyed, and persist for no time at all. And so from these considerations, belief might come along for the other cases, for it is natural that it should hold in the same way for them all.

Chapter 8

And that it is possible for there to be an infinite motion which is one and continuous, and that this is the one in a circle, let us now explain. For everything that *30* changes place is moved in either a circle, a straight line, or a mixture of the two, so that if one of those is not continuous, neither can that composed of both be continuous; and that something carried along a line that is straight and finite is not carried continuously, is obvious. For it turns back, and what turns back on a straight line is moved in opposite motions, since with respect to place, up is opposite to down, forward to backward, and leftward to rightward, *262a* for these are oppositions of place. But what a motion is that is one and continuous was marked out before [228b, 2–3], that it is of one thing, in one time, and in respect to something unvarying in kind (for there were three things, the thing moved, such as a human being or a god, the when, namely the time, and third that in respect to which the thing is moved, which is either place, attribute, species, or size). But contraries differ in kind, and are not one, and the differences of place are the ones mentioned. And it is a sign that the motion from

A to B is contrary to that from B to A, that, if they happen at the same time, they check and stop *10* each other. And likewise with the circle, the motion, say, from A toward B is contrary to that from A toward C (for they stop each other, even if each is continuous and no turning back takes place, because contraries destroy or prevent each other), but motion to the side is not contrary to upward motion.

But it is most clear of all that motion in a straight line cannot be continuous because what turns back must stop, and not only on a straight line but even if it is moved on a circle. For it is not the same thing to be moved in a circle and on a circle; for it is possible sometimes to go on being moved, sometimes to turn back again upon coming to the same place from which it set out. The belief that it must stop rests not only on the senses but also *20* on reason. The source of it is this. If there are three things, a beginning, a middle, and an end, the middle is both a beginning and an end in relation to the other two, and though one in number is two in meaning. Furthermore, what *is* potentially and what *is* actively are different, so that, of the points between the extremities of a straight line, any one whatever is potentially a middle, but is not so actively unless something cuts the line in two at that point and, having halted, begins to move again; and in this way the middle becomes a beginning and an end, a beginning of the later part and an end of the first. (I mean, for instance, if A, being carried along, stops at B and is again carried on to C.) But whenever it is carried along continuously, *30* A cannot have come to be at nor have departed from the point B, but can only be there in a now, but not for any time except within that of which the now is a division, within the whole. (And if anyone harbors the thought that it has come to be there and departed from *262b* it, A would always be stopping while being carried along, for it is impossible for A to have come to be at B and departed from it simultaneously. Therefore it did the one at one point in time, and the other at another. Therefore there would be a time in the middle. And so A would be at rest at B. But similarly also with the other points, for it is the same argument with them all. So whenever A, being carried along, uses the middle point B as both an end and a beginning, it must stop in order to make it two, just as one might also single it out in thinking.) But it departed from the starting point A, and came to be at the point C, when it finished and stopped.

10 For this reason too, the following must be said about the impasse, and the impasse is this: If the line E is equal to F, and A is carried continuously from the end of E toward C, while at the same time that A is at point B, D is being carried from the end of F toward H, uniformly and with the same speed as A, D would come to H before A came to C, for what is set in motion and sets out earlier must arrive earlier. For A did not come to be at B and depart from it at the same time, for which reason it came later. If it did both at the same time, it would not come later, but it *must* stop. Therefore one must not decide that at a certain time A came to be at B, at the same time that D was being moved from the end *20* of F (for if A were to have arrived at B, there would also be a departure from it, and not at the same time), but that it was there not in a time but in a cut in time. In this case, then, with something continuous, it is impossible to speak in that way, but with something that turns back it is necessary to speak that way. For if H were carried to D and, turning back again, were carried down its course, it would have used the endpoint D as an end and a beginning, using the one point as two; for this reason it must stop, and not simultaneously have come to be at D and have departed from D, since it would be there and not be there in the same now. And one may not invoke the previous solution; for it is not possible to say *30* that H is at D in a cut in time, but has not come to be at it or departed from it. For it is necessary to arrive at an end that is actively there, not potentially. So while the points in *263a* the middle *are* potentially, this one *is* actively, and from below is an end, from above a beginning; and it therefore belongs in the same way to the motions. It is therefore necessary for something that turns back on a straight line to stop. It is therefore not possible for a motion in a straight line to be everlastingly continuous.

And in the same way one must also meet those who ask the question that comes from Zeno's argument, whether it is necessary always to have come to the half-way point, though the half-way points are infinite and it is impossible to go through infinitely many things, or in the other way that some ask the question arising out of this same argument, demanding that, at the time something is being moved, one count the preceding half-motion resulting from each *10* half-distance that has come about, so that it follows that when the thing has gone through the whole motion, one has counted an infinite number, which is by agreement impossible.

Now in our first discussions about motion, we resolved this diffculty by way of time's having an infinity in itself [232b, 20–233a, 32; 239b,

5–240a, 18], for it is in no way absurd if someone goes through infinitely many things in an infinite time, and the infinite is present in the same way in both the distance and the time. Though this solution is sufficient against the one who raises the question (since he asked whether it is possible in a finite time to go through or count infinitely many things), in relation to the thing and the truth it is not sufficient; for if someone, leaving aside the *20* distance and the question whether it is possible in a finite time to go through infinitely many things, were to inquire of these same things about the time itself (since time has infinitely many divisions), this solution will no longer be sufficient, but one must articulate the truth, which is the thing we were saying in the passage just above. For if someone divides what is continuous into two halves, he uses one point as two, since he makes it a beginning and an end. And thus do both the one who counts and the one who divides into halves. But having been thus divided, neither the line nor the motion will be continuous; for a continuous motion is through something continuous, and while infinitely many halves are present in what is continuous, they are present not actively-and-completely but potentially. *30* And if one were to make them actively-and-completely be, he would make the thing not be continuous, and stop the motion, and it is clear that this very thing happens in the case of *263b* the one who counts the halves; for it is necessary for him to count the one point as two, since of one half it will be an end, and of the other a beginning, whenever one counts the continuous not as one but as two halves. So to the one who asks whether it is possible to go through infinitely many things either in time or in distance, one must say that in a certain way it is, but in another way it is not. For if things are infinitely many actively-and-completely, it is not possible, but it is possible if they are so potentially. For what is moved continuously has gone through infinitely many things incidentally, not simply; for it is incidental to the line to be infinitely many halves, but the thinghood of the line, and the being of it, are different from that.

10 And it is clear too that unless one makes the dividing point of before and after in the time belong to the later time for the thing, the same thing will at once be and not be, and will have come into being at a time when it is not. Now the point belongs in common to what is before and what is after it, and is the same and one in number, but it is not the same in meaning (since it is the end of one, the beginning of the other); but for the thing it always belongs to the later attribute. Call the time ACB and the thing D. This is white during time A, but not white during time B; therefore at time C it is white and not white. For in any part whatever of A it is true to say that it is white, if during this whole time it was white, and *20* during B it is not white; but C is in both. Therefore one must not grant the "during all . . ." but must leave out the last now, which is C; but this already belongs to the later time. And if not-white was coming into being and white dying away during all of time A, then at C the one has come into being and the other has been destroyed. So at that time it is first true to say that it is not white (or white [in the case of the opposite change]), or else something will not be when it has come into being, or will be when it has been destroyed, or will have to be white and not white at the same time, or generally be and not be. But if what is, and before was not, must come to be a being, and while it is becoming *is* not, it is not possible *30* to divide time into indivisible times. For if at time A, D was becoming white, but at a different indivisible time B next to it, it has become and at the same time is white—if at A *264a* it was becoming, but was not, but at B it is—then there must be a coming-into-being between them, and so also a time in which it was coming into being. For there will not be the same argument for those who deny indivisibles, but it came into being and is at the last point of the time itself in which it was coming into being, which nothing is next to or next in series with; but indivisible times are in series. But it is clear that if it was coming into being during the whole time A, there is no more time in which it has become or did become than that during all of which it was only coming into being.

These and similar arguments are some of those by which, as proper to the topic, one *10* might be persuaded; and it would also appear so to those who examine the topic logically, because the same thing follows. For everything that is moved continuously, if it is not bumped aside by anything, comes in the course of its change of place to the very place to which it was being carried beforehand; for instance, if it comes to B, it was also being carried to B, and not at the time when it was near it but straight from when it was first moved. For why now rather than earlier? And it is similar with other things. But say the thing carried from A, when it comes to C, will come back to A, while being moved continuously. Therefore, at the time when it is being carried from A to C, it is at that time also being carried in the motion to A from C, so that it is carried in contrary motions at the same time, since on a straight line they are contraries. Also, it is at the same time changing out of that *20* in which it is not. So if this is impossible, it must stop at C. Therefore the motion is not one; for a motion that is cut off by a stop is not one. And besides, from the following this will be more clear in general for every motion. For if everything that moves is moved with one of the kinds of motion mentioned, and rests with the state of rest opposite to it (for there was no other besides these), and if what is not always moved with this same motion (by which I mean all those distinct in kind, and do not mean something if it is part of the whole) must have been at rest beforehand with the opposite state of rest (since rest is the deprivation of motion); then if there are contrary motions on a straight line, and it is not possible to be *30* moved in contrary ways at the same time, the thing being carried from A toward C could not at the same time also be carried from C toward A, and since it is not being carried at the same time, but will be moved with this motion, it is necessary for it to rest beforehand *264b* at C, since this is the state of rest opposite to the motion from C. It is clear, then, from what has been said that the motion will not be continuous. And there is also the following argument, more native to the topic than what has been said. For the not-white has been destroyed at the same time that the white has come into being. Then if the alteration to white were continuous with that from white, and did not pause for some time, at the same time the not-white would have been destroyed, the white have come into being, and the not-white have come on the scene, for the same time would belong to the three. Further, it is not the case that if the time is continuous, the motion is too, but only that it is an unbroken series. And how could the same extremity belong to contraries, such as whiteness and blackness?

10 But motion on a circle will be one and continuous, since nothing impossible follows from that; for the thing moving from A will at the same time, by the same supposition, be moving to A (for that to which it comes is that to which it is moving), but it will not at the same time be moving in contrary or opposite ways. For not every motion to something is contrary or opposite to one from it, but is contrary if it is on a straight line (for there are on this contraries with respect to place, such as on a diameter, since they are separated the most), or opposite if it is on the same line. So nothing prevents a thing from being moved continuously with no time being left as a gap, since motion in a circle is from something itself to itself, while that on a straight line is from something itself to something else, and *20* that in a circle is never between the same things, while that on a straight line is repeatedly between the same things. Now that which is always coming to be in another and another place admits of being moved continuously, but that which is between the same places over and over does not admit of it, since it would have to be moved in opposite ways at the same time. Therefore neither in a semicircle nor in any other circumference is it possible to be moved continuously, since these would need to be moved over and over and to change in contrary changes, since the end does not join up with the beginning. But those of the circle do join up, and only it is complete.

And it is clear from the same distinction that the other kinds of motion cannot be continuous either; for in all of them the same things turn out to be moved through over and over, such as the in-between in alteration, or in change of the how-much, the sizes in the *265a* middle, and in coming into being and destruction something similar. For it makes no difference to make the things through which the change is be few or many, or to put something in between or take something away, for both ways it turns out that the same things are moved over and over. So it is clear from these things that the writers on nature do not speak well, who say that all sensible things are always in motion; for they must be moved with one of these motions, and most of all, according to these people, must be altering, since they say things are always in flux and wasting away, and also say that coming into being and destruction are alterations. But the present argument has said in general about every motion that it is possible to be moved continuously with no motion at *10* all apart from that in a circle, so that neither alteration nor increase can be continuous. To the effect, then, that there is no change that is infinite or continuous apart from change of place in a circle, let so much have been said by us.

Chapter 9

That circular motion is the primary kind of change of place is evident. For every change of place, as we said before, is either in a circle, on a straight line, or a mixture of the two. And the former two must be more primary than the last kind, since it is composed of them. And that in a circle is more primary than the straight, since it is more simple and complete. For it is not possible to be carried through an infinite straight line (for there is not anything infinite in that way, and even if there were, nothing would be moved *20* through it, since the impossible does not happen, and to go through the infinite is impossible). But motion on a finite straight line, if it turns back on itself, is composite and two motions, but if it does not turn back, is incomplete and destructible. But by nature, by definition, and in time, the complete is more primary than the incomplete, and the indestructible than the destructible. Besides, what admits of being everlasting is more primary than what does not admit of it; and while that in a circle admits of being everlasting, no other change of place nor any other of the rest of the motions does, for each must come to a stop, but if there is a stop, the motion is destroyed.

And it is reasonable that it has turned out that motion in a circle is one and continuous, and that in a straight line is not. For of motion on a straight line, a beginning, *30* middle, and end are marked out, and it has them all within itself, so that there is a place from which the moving thing begins and at which it will end (for everything that is at its limits, whether from which or to which, is at rest), but of the circumference, the beginning, middle, and end are indeterminate; for why would any sort of points on the line be more of *265b* a limit? Each of them is alike a beginning, a middle, and an end, so that the moving thing is both always and never at a beginning and at an end. For this reason the sphere is in a certain way both moving and at rest, for it holds the same place. And the cause why all these things happen is in the center; for it is both a beginning and a middle of the magnitude, and an end as well, and so, since it is off the circumference, there is no place where the thing carried along will come to rest as it goes through the circumference (for it is always carried around the middle, but not to the extremity), and since this stays still, the whole is in a certain way both always at rest and continuously in motion. And it follows *10* reciprocally, both because circular motion is the measure of motions, that it must be primary (for all things are measured by the first thing), and because it is primary, that it is the measure of the rest. And besides, only circular motion admits of being uniform; for those on straight lines are carried non-uniformly from the beginning and to the end, since everything is carried faster according to how much it departs from being a thing at rest, but of the circle alone there is neither beginning nor end within it by nature, but they are outside it.

But that change of place is primary among the kinds of motion, everyone bears witness who has made any mention about motion; for they give over the origins of it to *20* things that cause such a motion. For separation and combination are motions with respect to place, and love and strife move things in that way, since one of them separates things and the other combines them. And Anaxagoras too says that intelligence separates things as the first mover. And similarly, those who speak of no such cause, but say things are moved because of the void, even they say that the motion with respect to place is what things are moved in by nature (for the motion on account of the void is change of place and as if in a place), and they suppose that none of the other motions belong to the primary beings, but only to the things made of these, for they say that being increased or decreased or altered *30* belongs to things combined or separated out of the atomic bodies. And in the same way, all those who make coming into being and destruction be on account of density and rarity set these under control by combination and separation. And still alongside these are those who make the soul responsible for motion; for they say that which itself moves itself is the *266a* ruling origin of moving things, but an animal and everything with a soul moves itself with a motion in respect to place. And we say that only the thing moved with respect to place is moved in the governing sense; and if it is at rest in the same place, but increases or decreases or happens to alter, it is moved in some respect, but we do not say that it is moved simply.

That, then, motion always was and will be through all time, and what is the source of the everlasting motion, and also what motion is primary, and what motion alone admits of being everlasting, and that the first mover is motionless, have been said.

Chapter 10

10 That the first mover must be without parts and must have no magnitude, let us now explain, first drawing some distinctions about things that come before that conclusion. One of these is that it is not possible for anything finite to cause a motion lasting an infinite time. For there are three things, the mover, the moved, and third that in which the thing is moved, the time. And these are either all infinite, or all finite, or some of them—two or one—are finite. Let A be the mover, B the thing moved, and C an infinite time. Now let D move some part of B, and call the part E, but it will not move it for a time equal to C, since what causes motion for a longer time is a greater mover. So time F is not infinite. *20* So in this way I will use up A by adding to D, and B by adding to E, but I will not use up the time by always subtracting an equal amount, since it is infinite; and so all of A will move the whole B in a finite time within C. Therefore it is not possible to be moved by something finite with any infinite motion.

That, then, it is not possible for a finite thing to cause motion for an infinite time, is clear; and that it is impossible in general for an infinite power to be in a finite magnitude, is evident from the following. For let the greater power always be the one that does something equal in less time, such as heating or sweetening or throwing, or generally, causing motion. Therefore, the thing that is acted on must be affected in some way by the thing that *30* is finite but has an infinite power, and more by it than by anything else, since the infinite is greater than anything else. But surely the time would not admit of being anything at all. For if A is the time in which the infinite strength heated or pushed something, and AB the *266b* time in which some finite strength did, then by continually taking finite strengths in greater ratios to that one, I will at some time come to that which has caused the motion in the time A. For by continually adding to something finite I will exceed any definite amount, and by subtracting, likewise fall short. Therefore the finite power will move something in a time equal to that taken by the infinite one. But this is impossible; therefore nothing finite admits of having an infinite power.

But neither can there be a finite power in something infinite; even though there can be a greater power in a lesser magnitude, there will still be more in a greater one than that. *10* So let AB be infinite. Now BC has some power, which in a certain amount of time moved D; call the time EF. Now if I take the double of BC, it will move it in half of time EF (for let this be the proportion), and so will move it in the time FG. So then continually taking double magnitudes in this way, I will never go through AB, but will always get a time less than the one given. Therefore the power of AB will be infinite, since it exceeds every finite power, so long as for every finite power the time too must be finite (for if in a certain time, just so much power moves it, then a greater power will move it in a time that is less but *20* definite, following the reciprocal proportion); and any power, like any multitude or magnitude, that exceeds every definite amount, is infinite. But there is also the following way to show this: for we will take some power of the same kind as the one in the infinite magnitude, though it is in a finite magnitude, which will measure off the finite power in the infinite one.

That, then, it is impossible for there to be an infinite power in a finite magnitude, or a finite power in an infinite one, is clear from these things. But about things that change place, it would be good first to raise a certain impasse. For if everything that moves is *30* moved by something, how is it that some of those that do not themselves move themselves are moved continuously when they are not touching the mover, such as things that are thrown? And if the mover at the same time also causes something else to move, such as the air, which when it is moved causes motion, it is similarly impossible, once *it* is not touching *267a* the thing that first moved it, for it to be moved either, but all of them must be moved at the same time and have stopped whenever the first mover stops, even if it, as does the magnet, makes what it moves able to cause motion. But it is necessary to say this, that the thing first causing the motion makes the air or the water or some other thing that is naturally such as to be able to cause motion as well as be moved *be* able to cause motion; but it does not stop causing motion at the same time it stops being moved, but stops being moved at the same time that its mover stops moving it, and yet is still a cause of motion. And for this reason it moves something else next to it; and it is the same story with this. But it comes to an end when a lesser power of causing motion continually becomes present in the next thing. And *10* that comes to a final stop when the preceding mover no longer makes it able to cause motion, but only makes it be moved. And these must stop at the same time, the mover and the moved, as well as the whole motion. So this sort of motion comes about in things that admit of sometimes being moved and sometimes being at rest, and is not continuous, though it seems to be, since it belongs to things that are either in a row or touching, for the thing causing the motion is not one, but a series of things next to each other. This is why such a motion happens in air or water, the kind of motion that some people say is circular replacement. But it is impossible to resolve the things raised as impasses except in the way described. Circular replacement makes everything be moved and cause motion at the *20* same time, and so also stop at the same time; but the present example appears to be some one thing being moved continuously. By what, then? For it is not moved continuously by the same thing.

But since there must be among beings a continuous motion, and this is one, and this one motion must belong to a magnitude (for what is without magnitude does not move), and must belong to one thing and be caused by one thing (otherwise it would not be continuous but one next to and divided from another), then if the mover is one, it causes motion either *267b* while being moved or while being motionless. But if it is moved, it would need to follow along and itself be changing, and at the same time be moved by something, and so the series will stop and come to what is moved by something motionless. For this need not change along with what it moves, but will always be able to cause the motion (since what causes motion in this way is without exertion), and this motion, either alone or most of all, is uniform, since what is moving it has no change at all. But the thing moved must not have any change either in relation to it, in order that the motion be uniform. So what moves it must be either at the center or on the circle, since these are sources of it. But the things closest to the mover are fastest; and the motion of the circle is such, so the mover is there.

10 But there is an impasse if it is possible to cause some moving thing to move continuously, but not in the manner of something that pushes it again and again, so as to be continuous by means of being one after another; for either the mover itself must always be pushing or pulling or both, or else there is some other thing taking it up one after another, as was said before about things thrown, if the air is divisible and another part repeatedly being moved causes the motion. But both ways it would be impossible for the motion to be one rather than consecutive. Therefore that alone is continuous which something motionless causes to move; for it is always in the same condition, and so will be related continuously in the same way to the moving thing.

And now that these things have been distinguished, it is clear that it is impossible for *20* the first and motionless mover to have any magnitude. For if it had magnitude, it would have to be either finite or infinite. But that it is not possible for a magnitude to be infinite has been shown earlier in the books about nature; and that something finite cannot have an infinite power, and also that it is impossible to be moved by anything finite for an infinite time, have been shown just now. But the first mover causes an everlasting motion and acts for an infinite time. It is clear, then, that it is indivisible and without parts, and has no magnitude.

Commentary on Book VIII, Chapters 7–10

The repeated narrowing of the most proper sense of motion, which began in Book V, which required gaining a sharper understanding of the continuity of motion, and which came to a conclusion in Book VII, is now seen as a whole. Not only are changes of thinghood and qualitative changes discontinuous in some respects, and not only is change of place a necessary condition for them and for growth or shrinking, but it is now observed that of the four kinds of change, change of place is a change least of all. The most continuous and primary kind of change is the one that tampers least with what the changing thing is, by changing only its place, its relation to things other than itself. Change and motion name the same events, but, like the road from Athens to Thebes and that from Thebes to Athens, they do not mean the same thing. The road from birth and death, to development and differentiation, to growth and wasting away, to change of place arrives at motion pure and simple, departing from change that is the most radical and complex. It is clear too that the road through the kinds of change goes from the way of being that belongs to living things, and to the way of being that belongs to the cosmos. Living and non-living nature form the two poles between which the ordered scale of motions ascends, while the same steps constitute a descent along the direction of change. But Aristotle is clear about which pole is at the top; the cosmos could continue with no living things in it, but life is impossible without conditions supplied by the being-at-work of the cosmos.

The largest part of this section is devoted to discovering the primary motion within change of place. Aristotle first narrows down the choices to motions along a straight line or a circle. This may appear to be an arbitrary restriction, but these are in fact the candidates that have appeared again and again throughout history as accounting for the ultimate motions. Ptolemy and Copernicus derive all motions from circles, Kepler from a combination of circles and one straight line, Galileo from straight lines and one circle, Descartes from straight lines that immediately produce whirlpools, and Newton from straight lines alone (but even he uses circles to measure the curvature of the motions compounded out of straight lines). No everlasting motion in one direction along a straight line is possible, since Aristotle has concluded that nothing in the world can be infinite in extent. If the primary motion is straight, it must perpetually reverse its direction. Aristotle argues at length that no reversal of a motion could take place instantaneously, but must take some time, at rest in the limiting position. At 263a, 3, he clearly has in mind a stone thrown straight up, that must stop for some duration of time before it begins to fall. His conclusion makes sense as a description of what we see, but it is not correct.

In his *Two New Sciences*, Galileo shows (National Edition pages 200–201), that the stone thrown straight up is at its peak for no duration, but only for an indivisible time. It is no different from a stone thrown in a gentle arc, that passes through the peak of its curve without stopping. Galileo founds his new science of motion on the consideration of projectiles, rejecting Aristotle's claim that there is anything un-natural about their motions. Even the course of the life of an animal or plant is nowadays sometimes described on the analogy of projectile motion; we are told that we are dying from the moment of birth, and visibly declining from the instant that catabolism overbalances anabolism. But it makes no sense to Aristotle to consider the mature form of a living thing a transient state; maturity is an active state of repose, an equilibrium that the living thing maintains, and its form governs the whole of its life. Aristotle and Galileo look in different places for the central image of nature. But even Galileo agrees that projectile motion is not elementary but composite. Aristotle's conclusion about reversing motion on a straight line does not depend at all on the claim that the change of direction takes time, but only on the recognition that such a motion is not single and continuous but needs two causes. The stone thrown upward rises because of the thrower, but falls because of its own heaviness. Since its reversing motion cannot have a single unvarying cause, it is not the primary motion, and that is all Aristotle takes from the consideration of it. And it is obvious that the motion of circular rotation is not composite but one, not broken but continuous, could have a single changeless cause, and, as the only changeless change, is the primary motion.

The long argument that says a reversing motion must stop for some amount of time, though it is incorrect and unnecessary to Aristotle's purpose, does have an important by-product that is entirely correct and central to Aristotle's purpose, and this is his final refutation of Zeno. In Book VI Aristotle showed that the paradoxes resulting from the infinite divisibility of distance dissolve when one notices that time is infinitely divisible in just the same way, so that the two match up. But what if someone asked how time itself can be gone through, if it has an infinity of divisions? To meet this difficulty, Aristotle must discuss Zeno's mistake in the deepest way, and this depends on the distinction that Aristotle used as the foundation of his definition of motion, that between potency and being at work. A magnitude, a motion, and a time, because they are continuous, are infinitely divisible but not infinitely divided. To count the divisions in something continuous means to take each one out of potency into the being at work of counting, and in this way to make a stop at each one. In a continuous motion, the moving thing is not arriving at and departing from each of its divisions, but is at each, as Aristotle says (262b, 22), not in a time but in a cut in time. To be continuous is, incidentally, to be capable of an infinity of divisions, but this is not what defines the continuous thing. In Books V and VI, Aristotle got the definition of continuity straight, seeing that what holds the continuous thing together takes precedence over its capacity to be torn apart. And in the case of motion, the definition of motion tells us that it is the prior existence of potencies in beings that holds each motion together as one and continuous. We can now see that Zeno's fundamental error is in failing to recognize anything as governing the wholeness of a motion. For Zeno, the motion is a collection of isolated, externally related, transient states. From that starting point alone, there is no way to put the pieces together into a motion, and Zeno's conclusion that there is no motion just makes explicit what is already built into his assumptions.

Aristotle's argument takes up again with the conclusion that only circular motion can be both whole and everlasting. The digression into the final refutation of Zeno has the effect of gathering all that has gone before in the *Physics*, from Book I on, into that conclusion, as it is taken forward to the last step of the inquiry. And the last question is, where does the power for this infinite motion come from? No finite mover has an infinite power, and no infinite magnitude can exist all at once. The difficulty could be met by Galileo or Newton, or by Philoponus before them, in the way discussed above in the commentary on the section on the void. If a finite power could be impressed upon a body, and just stay there on top of its nature, overpowering the body's own being-at-work with an inert impulse, then the effects could just go on forever, and everyone who throws a rock would be a perpetual-motion machine. But projectile motion does obviously go on after the body separates from the thrower, and this is a difficulty Aristotle must meet. If there can be no inertia, and the stone flies for awhile after it leaves the impelling hand, some other cause of motion must have taken over from the hand the responsibility for the stone's continuing to move. Aristotle suggests that the air next to the thrown stone is not only moved but energized, as a magnet energizes a piece of iron, and temporarily sustains the stone and pushes it on, energizing in turn another portion of the air, though more weakly, until some portion is no longer powerful enough to move the stone any farther. This explanation is much ridiculed (though it is hard to see that there is anything less acceptable about it than about a gravitational or electromagnetic field, which must do all the tricks Aristotle needs the air to do without the advantage of being corporeal), but Aristotle has no positive interest in projectile motion, which is constrained, derivative, composite, and incidental. His only concern here is to show that it cannot be a model for the action of the first mover.

Aristotle's conclusion is that the first mover can have no magnitude at all. It must be a source outside of nature, in contact with the outermost sphere of the cosmos, responsible by its own constant being-at-work for the unvarying first motion that holds together the world. This is the end of one long dialectical ascent, which begins in Book I with the affirmation that there is change and with the question of what sort of thing would be able to undergo change. The sources of change are found in the forms of living things, and in the one motionless first mover of the cosmos. What the latter might be is a question that does not belong to physics, but that question converges with a whole array of other questions uncovered at the ends of other inquiries. The *Nicomachean Ethics* discovers the highest source of human happiness in an activity of contemplation. In *De Anima* it is discovered that the highest condition of human thinking is an active intellect. All these inquiries merge into that in the *Metaphysics*, which asks what is ultimately the source of the thinghood of things. All the things one might study are bound up together, and point to the source of all being. Physics discovers, and abandons itself on, the doorstep of first philosophy.

Appendix

Digressive Chapters

Book I, Chapter 3 (On Parmenides and Melissus)

For those who approach the topic in this way, it seems impossible that beings are one, and the things out of which they demonstrate it are not difficult to refute. For both Melissus and Parmenides reason *10* like debaters. That Melissus mis-reasons is obvious, for he supposes it to have been given, if everything that comes into being has a source, that therefore what does not come into being has none. And this too is absurd, that there must be of everything a source of the thing rather than a beginning of the time, and not of coming into being simply, but also of alteration, as though something could not change as a whole. Then too, why is something motionless if it is one? For just as a part which is one, the water around us, say, is in motion within itself, why not also the whole? And on top of that, why would there not be alteration? And surely it is not *20* possible for the whole to be one in form, but only in what it is made out of (and even some of those who study nature say it is one in this way, but not in the other), for a human being is different in form from a horse, and opposites from one another.

On the side of Parmenides there are arguments of the same kind, even if there are some others peculiar to him. And the refutation in one way is that it is false and in another that it does not follow: false in that he takes being to be meant simply when it is meant in more than one way, but not a necessary conclusion anyway because, supposing only white things were taken, white meaning one thing, nonetheless white things are many and not one. For neither by continuity nor by meaning will what is white be one, since being white is different *30* from being receptive of whiteness. Nor will what is receptive of it be separate, apart from the whiteness, for not as something separate but in the being of it is whiteness different from that to which it belongs. But Parmenides did not share in this observation.

Accordingly, he had to take being not only to mean one thing, in whatever respect it was attributed, but also to mean the very thing which is and is one. For the attribute is predicated of some underlying thing, so that *186b* to which *being* is attributed would not *be* (since it is other than being), and there would therefore be something which is not. So there could be no other thing to which the very thing which is belongs, since there would not be any being it could be, unless being were to mean many things in such a way that each could be something. But it is laid down that being means one thing. If, then, the very thing which is can be attributed to nothing, why would it signify being rather than not-being? For if the very thing which is were also white, though being white is not the very thing which is (for it is impossible to attribute being to it since nothing is a being which is not *10* the very thing which is), the white thing would therefore not be a being, and not in the sense of not being something in particular, but of not being at all. Therefore the very thing which is cannot be a being, since it is true to say that it is white, but this would mean not-a-being. So the white thing does also signify the very thing which is, and being therefore means more than one thing.

Then, too, being would not have magnitude, if being is the very thing which is, for the being of each of its parts would be different [from that of the whole]. But that the very thing which is must be divided into some other very things which are is clear from definitions; for example, if human being is the very thing which is, it is necessary that animal and two-footed also be very things which are. For if each is not a very thing which is, it would be an attribute, either in the human being or in some other underlying thing. But that is impossible, *20* since an attribute means this: either that which admits of belonging or not belonging to something, or that in the definition of which the thing to which it is attributed belongs. (For instance, sitting is a separable attribute, but in snubness there belongs the definition of the nose, to which we say it is attributed.) Yet whatever is present in the defining articulation of a thing, or of the things out of which it is made, does not have present in its own articulation that of the whole; as that of human being is not present in that of two-footed, or that of white human being in that of white. If, then, this is the way of these things, and two-footed is attributed to human being, it must be separable, so that a human being would admit of not being two-footed, or else the *30* articulation of human being would be present in that of two-footed, which is impossible, since the articulation of the latter is present in that of the former. But if two-footed is attributed to something else, and animal as well, and neither of these is the very thing which is, then also human being would be among the attributes of something else. But let it be that the very thing which is cannot be an attribute of anything, and let the composite of the two attributes be predicated of that to which both are attributed: is the whole therefore composed of indivisibles?

187a Some people gave in to both arguments: to the argument that all things are one, if being means one thing, by accepting that there is non-being, and to the one from cutting in half by making atomic magnitudes. But it is clear that it is not true, because if being means one thing and cannot at the same time be its contradictory, there will not be any non-being; for it does not at all follow that anything simply is not, but that what is not is not something in particular. And to say that if there were not anything apart from being itself, all things would be one, is absurd. For who understands being itself other than as being some very thing that *10* is? But if this is so, still nothing prevents things from being many, as was said. That, then, it is impossible for being to be one in this way, is clear.

Book I, Chapter 4 (On Anaxagoras)

But there are two ways in which those who study nature speak. For some, making the underlying body one, either one of the three or something else which is denser than fire but more rarefied than air, generate the other things by a process of becoming more dense or more rare, thus making things many. (And these are contraries, in the general class of being more and being less, just as Plato speaks of the great and the small, except that he makes these material and makes the one their form, while they make the underlying material one *20* and make the contraries be distinctions and forms.) But the others make the contraries separate themselves out from the one thing in which they inhere, as Anaximander says, as well as all those who say things are both one and many, as Empedocles and Anaxagoras do, for they also separate out the other things from what is mixed. But they differ from one another in that one makes these things happen in a cycle, the other once only, and again in that one makes them infinite, both the homogeneous things and the contraries, the other the so-called elements only.

It is likely that Anaxagoras supposed things thus to be infinite because he assumed the common opinion of those who study nature to be true, that nothing comes into being from what is not. (For on this account they *30* make statements of the kind "All things were together," and set it down that to come into being is to alter in a certain way, by a process of combining and separating according to some.) Another reason is from the coming into being of opposites from one another; they were therefore already present, for if everything that comes into being must come either out of what is or what is not, and of these, the coming out of what is not is impossible (for about this opinion, all those who concern themselves with nature think alike), they regarded the remaining choice as following immediately by necessity, that coming into being is from what is and is already *187b* present all along, but from things imperceptible to us on account of the smallness of their bulk. For this reason they say everything is mingled with everything, because of seeing everything coming into being out of everything, but things appear different and are named differently from one another from what predominates most in multitude in the mixture of an infinity of things, since there is no wholly unmixed white or black or sweet or flesh or bone, but whatever each thing has most of, this seems to be the thing's nature.

Now if the limitless as limitless is unknowable, the infinite in multitude or magnitude is unknowable *10* as to how much it is, and the unlimited in form is unknowable as to what kind of thing it is. But since the original beings are infinite both in multitude and in kind, it is impossible to know the things made of them. For we assume that we know a composite thing in this way: whenever we know how many things it is made of and what they are. Further, if necessarily that of which the part can be of any size whatever in greatness or smallness can also itself be of any size (and I take such parts to be whatever, being already present all along, the whole is divisible into), but if it is impossible that an animal or plant be of any size whatever in greatness or smallness, then it is clear that neither can any of its parts, since the whole would also be the same way. But flesh and bone *20* and such things are parts of an animal, and fruit of plants. It is obvious then that it is impossible for either flesh or bone or anything else to take on any size whatever in magnitude either in the direction of the greater or that of the less. Further, if all such things inhere all along in one another, and do not come into being but, being present, separate themselves out, and are spoken of from what is predominant in them, and anything at all comes out of anything (such as water separated out of flesh and flesh out of water), and in every case a finite body is taken away from a finite body, it is clear that each thing does not admit of being present all along in each thing. For when flesh has been taken away out of the water, and again something else has come out of the remainder *30* by separation, even if what is separated out were always less, still it could not go beyond a certain magnitude in smallness. So if the process of separating out were to come to an end, not everything would be present in everything (for in the remaining water, flesh would not inhere), but if it did not come to an end, but always had a process of taking away, in a finite magnitude there would be present equal finite things infinite in multitude, which is impossible. And in addition to these things, if every body, when something is taken away from it, necessarily becomes less, and the how-much of flesh has a limit in both greatness and smallness, it is clear that *188a* from the least amount of flesh no body could be separated out, since it would be less than the least. Further, among the infinite bodies, there would be present already infinite flesh and blood and brain, separated from one another but being what they are nonetheless, and each one infinite; but this is irrational.

But that things are never completely separated, though not knowing he is saying it, he asserts rightly. For attributes cannot stand as separate. So if colors and states are mixed, if they were to be separated out there would be *10* something white or healthy which was not anything else and did not belong to an underlying thing. So the intelligence [Anaxagoras speaks of] is absurd, since it seeks what is impossible, if indeed it wants to separate them; and this is to do the impossible both in the case of the how-much and in that of the of-what-kind, the former because there is no least magnitude and the latter because attributes cannot be separate. Nor does he rightly grasp the coming into being of homogeneous things. For there is a way in which clay is divided into clay, and a way in which it is not. And it is not in the same manner as bricks come from a house and a house from bricks that water and air both are and come to be from one another. And it would be better to have taken things as fewer and finite, as Empedocles does.

Book V, Chapter 5 (Motions as Opposite to Motions)

It is still necessary to distinguish what sort of motion is contrary to a motion, and in the same way concerning rest. And one must first distinguish whether a motion from *10* something is contrary to the one to the same thing (such as one from health to one to health), as also coming into being and destruction seem to be, or whether the contrary motions are the ones from contraries (as one from health to one from sickness), or to contraries (as one to health to one to sickness), or the one from one contrary to the one to its contrary (as one from health to one to sickness), or the one from one contrary to its contrary to the one from the other contrary to its contrary (as one from health to sickness to one from sickness to health). For it must be in one or more of these ways, since there is no other way of being opposite. But the one from a contrary is not contrary to the one to its contrary, as one from health to one to sickness, since they are one and the same. The being of them, however, *20* is not the same, just as it is not the same to change from health as to change to sickness. Nor is the one from a contrary the contrary of the one from its contrary, for it happens at the same time that what is from a contrary is also to either its contrary or something in between, but we will speak about this later; but it would seem that the change to a contrary, rather than that from a contrary, is the cause of contrariety. For the latter is a departure from contrariness, but the former is a taking hold of it. Also, each motion is described by that to which, rather than that from which, it changes, as healing is the motion to health, getting sick that to sickness.

So there remain the motion to a contrary, and that to a contrary from its contrary. *30* Now it may incidentally be that those to contraries are also from contraries, but still the being of them is not the same; I mean to-health is not the same as from-sickness, nor from-health as to-sickness. But since change differs from motion (motion being the change from *229b* a subject to a subject), the motion from one contrary to its contrary is the contrary of that from the other contrary to its contrary, as one from health to sickness to one from sickness to health. And it is clear from looking at examples what sorts of things seem to be contraries: getting sick to getting well, or learning to being deceived by someone other than oneself. (For these are into contraries; for just as with knowledge, there is also deception acquired both through oneself and through someone else.) Other examples are change of place upward to that downward (for these are contrary as to height), or that rightward to *10* that leftward (for these are contrary as to breadth), or that forward to that backward (for these too are contrary). That which is into a contrary only is not motion but change, such as coming into being-white, not from anything. And with however many things there are that have no contraries, change from them is contrary to change to them; hence coming into being is contrary to destruction, and loss to gain, but these are changes, not motions. And with however many contraries there are that have in-betweens, one must set down motions to what is in between as in a way to contraries; for the motion uses the in-between as a contrary on whichever side it changes. For instance, a motion from gray to white, is as from black, from white to gray, as to black, or from black to gray as to white, the gray taking *20* either place. For the middle is said to be in a way either of the extremes in relation to the other, as was said before. A motion is contrary to a motion then in this way: that from one contrary to its contrary to that from the other contrary to its contrary.

Book V, Chapter 6 (Motions as Opposite to Rest)

But since to a motion not only a motion but also rest seem to be contrary, this is something that must be distinguished. For what is simply contrary to a motion is a motion, but rest too is opposite to it (for it is a deprivation, and there is a way in which the deprivation is also said to be contrary), and a certain kind of rest to a certain kind of motion, as that with respect to place to that with respect to place. But this is a simple way of saying it; for is it the motion from this place or to it that is opposite to rest there? But *30* it is clear that, since motion belongs to subjects in two ways, the one *from* this place to its contrary is contrary to the rest there, and the one from the contrary place to this one to the rest in the contrary place. And at the same time, these are also contrary to one another; *230a* for it would be absurd if motions were contrary but states of rest were not opposites. And they are the ones in contrary states, as rest in health to that in sickness. (And it is contrary to the motion from health to sickness, for it is unreasonable that rest in health be contrary to motion from sickness to health, since the motion into that in which it stands still is rather the coming to rest, or at least it happens to come into being at the same time as the motion, and the contrary motion must be one of these two.) Certainly rest in whiteness is not contrary to rest in health. And with however many things there are that have no contraries, the change from one of them is opposite to the change to it, but is not a motion, examples being *10* change out of or into being, and with these there is no state of rest, but of being-unchanged. And if there should be some particular subject, its being-unchanged in its being is contrary to being so in its not-being. And if it were not possible for there to be some particular non-being, one would be at an impasse about what a thing's being-unchanged in its being was contrary to, and if it were a state of rest. But if it were this, either not every state of rest is contrary to a motion, or coming into being and destruction are motions. It is clear, then, that unless these are also motions, that state ought not to be called rest, but something similar, being-unchanged; and it is contrary either to nothing, or to being-unchanged in its not-being, or to its destruction, since this is from it, while coming into being is to it.

20 Now someone might be at an impasse why it is, in the case of change of place, that there are states of rest and motion both in accordance with and contrary to nature, but not in the other motions, such as alteration, is there a distinction of according to or contrary to nature (since neither getting well nor getting sick is any more according to or contrary to nature, nor turning white or turning black). And it is similar with increase and decrease (for neither are they opposed to one another as by and contrary to nature, nor is one increase so opposed to another increase). And it is the same story with coming into being and passing away, since neither is it the case that coming into being is by nature and passing away contrary to nature (for growing old is by nature), nor do we see coming into being sometimes according to and sometimes contrary to nature. But if what is by force is contrary *30* to nature, would not also a forcible destruction be opposed to another destruction as what is contrary to nature to what is in accordance with it? And are there not some comings into being that are constrained and not as they are meant to be, to which are opposed those *230b* according to nature, and forced increases and decreases, such as rapid weight-gain in the young on account of luxuriousness, or fast-ripening grain that is not dense-textured? And how is it with alteration? Is it not just the same? For some would be forced and others natural, as some people throw off disease on other than the critical days, others on the critical ones; the ones, then, altered contrary to nature, the others in accordance with it. And so destructions would be contrary to one another, not to comings into being. And what prevents it from being so in a certain way? For it would also be the case if one were pleasant and the other painful; so a destruction is not simply contrary to a destruction, but *10* insofar as one of them is such-and-such and the other so-and-so.

In general, then, motions and states of rest are contrary in the way that has been said, as are up and down, since these constitute a contrariety of place. And fire is carried in an upward motion by nature, and earth downward, and their changes of place are contrary. With fire the upward is by nature, the downward contrary to nature, and its motion according to nature is contrary to the one contrary to nature. And similarly with states of rest: for being at rest above is contrary to motion downward from above. But to earth that state of rest happens contrary to nature, and that motion by nature. So to a motion in *20* accordance with nature, a state of rest of the same thing contrary to nature is contrary, since also the motion of the same thing is contrary in that same way, for one of its motions, upward or downward, will be according to nature, the other contrary to nature.

But there is an impasse: whether of every non-eternal state of rest there is a coming into being, and whether this is a coming to a standstill. But then there would be a coming into being of a state of rest contrary to nature, as of earth above. Therefore when it was being carried upward by force, it would have been coming to a standstill. But what is coming to a standstill seems always to be speeding up, but what is moved by force seems to do just the opposite. Therefore, not having come to be in a state of rest, it will be in a state of rest. Further, coming to a standstill seems altogether to be, or to go along together with, a thing's being carried to its own place.

And there is an impasse if rest in a place is contrary to motion from that place; for *30* whenever a thing is moving away from or throwing off something, it seems still to have what is being thrown off, so if this state of rest is contrary to a motion from one place to its contrary, the contraries will belong at the same time to the resting thing. But is it not the *231a* case that, if it is still remaining in place, it is at rest in a certain way, but generally, part of the moving thing is there, part at that toward which it is changing? Hence also, a motion is contrary to a motion more-so than is a state of rest.

And about motion and rest, how each is one, and which ones are contraries, have been said.

Glossary

This glossary has three purposes: to be read through as a general orientation to Aristotle's thinking, for reference, and to point to those places in and outside the *Physics* in which Aristotle explains and clarifies his own usage. Bekker page numbers from 184 to 267 refer to the *Physics*; those from 980 to 1093 are in the *Metaphysics*.

abstraction *(aphairesis)* — The act by which mathematical things, and they alone, are artificially produced by taking away in thought the perceptible attributes of perceptible things (1061a, 28–b, 2). Within mathematics, this is the ordinary word for subtraction. It is never used by Aristotle to apply to the way general ideas arise out of sensible particulars, as Thomas Aquinas and others claim. Its special philosophic sense is not Aristotle's invention; as often as not he speaks of "so-called abstractions." He uses the word in this special sense rarely, only in reference to the origin of mathematical ideas, and not always then; in the *Physics* he says instead that mathematicians separate what is not itself separate (193b, 31–35).

active state *(hexis)* — Any condition that a thing has by its own effort of holding on in a certain way. Examples are knowledge and all virtues or excellences, including those of the body such as health. Of four general kinds of qualities described in *Categories* VIII, these are the most stable.

alteration *(alloiōsis)* — Change of quality or sort, dependent upon but not reducible to change of place. One of the four main kinds of motion. Some things that we would consider qualities are "present in" the thinghood of a being, making it what it is, rather than attributes of it; change of any of them would be change of thinghood, rather than alteration of a persisting being (226a, 27–29). The acquisition of virtue is just such a change of thinghood, not an alteration but the completion of the coming-into-being of a human being, just as putting on the roof completes the coming-into-being of a house (246a, 17–246b, 3). For a different reason, learning is not an alteration of the learner; knowing is a being-at-work that is always going on in us, unnoticed until we settle into it out of distraction and disorder (247b, 17–18).

ambiguity *(homōnymia)*
The presence of more than one meaning in a word, sometimes by chance (as in "bark"), but more often by analogy or by derivation from one primary meaning. (See especially *Metaphysics* IV, 2.) A city or society is called healthy by analogy to an animal, a diet by derivation. Derived meanings may have many kinds of relation to the primary meaning, but all point to one thing (*pros hen*). Arrays of this truthful kind of ambiguity reflect causal structures in the world. Book V of the *Metaphysics,* mistakenly called a dictionary, is called by Aristotle the book about things meant in more than one way. Thomas Aquinas uses the word "analogy" to cover all non-chance ambiguity, but it makes a great difference to Aristotle that the meanings of *good* are unified only by analogy, while those of *being* point to one primary instance.

art *(technē)*
The know-how that permits any kind of skilled making, as by a carpenter or sculptor, or producing, as by a doctor or legislator. The artisan is not "creative"; in nature the form of the thing that comes into being is at work upon it directly, while in art the form is at work upon the soul of the artisan (1032b, 13–14). Aristotle agrees with sculptors that Hermes is in the marble, and is let out by taking away what obscures his image. Aristotle concludes that the origin of motion that produces statues is the art of sculpture, and

incidentally the particular sculptor (195a, 3–8). The artwork or artifact has no material cause proper to itself (192b, 18–19—though a saw needs to be of a certain kind of material to hold an edge); in general the artisan uses the potencies of natural materials to counteract one another. The surface of a table strains to fall to earth, but the legs prevent it, while the legs strain to fall over and the tabletop prevents it, and similarly with the roof and walls of a house.

articulation *(logos)*
: The gathering in speech of the intelligible structure of anything, a combination of analysis and synthesis. A definition is one kind of articulation, but there are many others, including a ratio, a pattern, or reason itself. It can refer to anything that can be put into words—an argument, an account, a discourse, a story—or to the words into which anything is put—a word, a sentence, a chapter, a book. Translating *logos* as formula is misleading, since it has no implication of being the briefest, or any rigid, formulation of anything. In some translations, the word "formula" becomes a formula for a rich and varied idea; the word "articulation" is a slight improvement, used here wherever nothing better was appropriate.

being-at-work *(energeia)*
: An ultimate idea, not definable by anything deeper or clearer, but grasped directly from examples, at a glance or by analogy (1048a, 35–37). Activity comes to sight first as motion, but Aristotle's central thought is that all being is being-at-work, and that anything inert would cease to be. The primary sense of the word belongs to activities that are not motions; examples of these are seeing, knowing, and happiness, each understood as an ongoing state that is complete at every instant, but the human being that can experience them is similarly a being-at-work, constituted by metabolism. Since the end and completion of any genuine being is its being-at-work, the meaning of the word converges (1047a, 30–31; 1050a, 21–23) with that of the following, *entelecheia*.

being-at-work-staying-itself *(entelecheia)*	A fusion of the idea of completeness with that of continuity or persistence. Aristotle invents the word by combining *enteles* (complete, full-grown) with *echein* (= *hexis,* to be a certain way by the continuing effort of holding on in that condition), while at the same time punning on *endelecheia* (persistence) by inserting *telos* (completion). This is a three-ring circus of a word, at the heart of everything in Aristotle's thinking, including the definition of motion. Its power to carry meaning depends on the working together of all the things Aristotle has packed into it. Some commentators explain it as meaning being-at-an-end, which misses the point entirely, and it is usually translated as "actuality," a word that refers to anything, however trivial, incidental, transient, or static, that happens to be the case, so that everything is lost in translation, just at the spot where understanding could begin.
cause *(aitia)*	The source of responsibility for anything. It thus differs in two ways from its prevalent current sense: in always being a source (1013a, 17) rather than the nearest agent or instrument that leads to a result, and in referring more to responsibility for a thing's being as it is than for its doing what it does. To understand anything is to know its cause, and such an understanding is always incomplete without an account of all four kinds of responsibility: as material, as form, as origin of motion, and as end or completion (*Physics* II, 3).
chance *(automaton)*	Any incidental cause. At 197b, 29–30, Aristotle invents the etymology *to auto matēn,* that which is itself in vain (but produces some other result). Chance events or products always come from the interference of two or more lines of causes; those prior causes always tend toward natural ends or human purposes. Chance is thus derivative from the "teleological" structure of the world, and is the reason nature acts for the most part, rather than always, in the same way. When chance is peculiarly relevant to a human being, it is called fortune or luck.

contradictory
(antiphasis)
One of a pair of opposites which can have nothing between them, such as white and not-white.

contrary
(enantion)
One of a pair of opposites which can have something between them, such as white and black; but the opposition need not be extreme, and could be between two shades of gray.

contemplation
(theōria)
The being-at-work of the intellect (*nous*), a thinking that is like seeing, complete at every instant. Our ordinary step-by-step thinking (*dianoia*) aims at a completion in contemplation, but it also presupposes an implicit contemplative activity that is always present in us unnoticed. To know is not to achieve something new, but to calm down out of the distractions of our native disorder, and settle into the contemplative relation to things that is already ours (247b, 17–18). An analogy to the relation between step-by-step reasoning and contemplation may be found in two ways of looking at a painting or a natural scene; one's eyes may first roam from part to part, making connections, but one may also take in the sight whole, drinking it in with the eyes. The intellect similarly becomes most active when it comes to rest.

deprivation
(sterēsis)
The absence in something of anything it might naturally have. Aristotle regards the distinction between the deprivation, which is opposite to form, and the material, which underlies and tends toward form, as a clarification of and advance over the opinions that came out of Plato's Academy (*Physics* I, 9).

dog days
(hupo kuna)
The hottest time of the year, when the dog-star, Sirius, first becomes visible above the horizon in the latitudes of Greece.

end
(telos)
The completion toward which anything tends, and for the sake of which it acts. In deliberate action it has the character of purpose, but in natural activity it refers to wholeness. Aristotle does not say that animals, plants, and the cosmos *have* purposes but that they *are* purposes, ends-in-themselves. Whether any of them is in another sense for the sake of anything outside itself is

always treated as problematic in the theoretical works (*Physics* 194a, 34–36; *Metaphysics* 1072b, 1–3; *De Anima* 415b, 2–3), though *Politics* 1257a, 15–22, treats all other species as being for the sake of humans. As a settled opinion found throughout his writings, Aristotle's "teleology" is nothing but his claim that all natural beings are self-maintaining wholes.

example *(epagōgē)*
The perceptible particular, in which the intelligible universal is always evident. The word induction, which refers to a generalization from many examples, does not catch Aristotle's meaning, which is a "being brought face-to-face with" the universal present in each single example. A famous simile in the last chapter of the *Posterior Analytics* (100a, 12–13) is often taken to mean that the universal must be built up out of particulars, just as a new position of a routed army is built up when many men have taken stands, but it means just the opposite: it only takes one man to take a stand, after which every other soldier, down to the original coward, will be identical to him. The rout corresponds to the condition of someone who has not yet experienced some universal in any of its instances. Evidence for this interpretation is found in many places, such as *Posterior Analytics* 71a, 7–9, and *Physics* 247b, 5–7, in which Aristotle unmistakably says that one particular is sufficient to make the universal known. That in turn is because the same form that is at work holding together the perceived thing is also at work on the soul of the perceiver (*De Anima* 424a, 18–19).

first philosophy *(prōtē philosophia)*
The study of immovable being, or of the sources and causes of all being. Aristotle's organized collection of writings on this topic was called by librarians *ta meta ta phusika,* "what comes after the study of natural things," but neither this phrase nor any like it is ever used by Aristotle. He names the topic in the order of the things themselves, rather than from the way we approach and think about them, and calls physics

second philosophy. What we call metaphysics, the post-natural, is for Aristotle the pre-natural, the source and foundation of motion and change. That form is present in all things is a starting-point for physics; what form is must be clarified by first philosophy (192a, 34–36).

form *(morphē or eidos)* — Being-at-work (1050b, 2–3). It is often said that Aristotle imports the form/material distinction from the realm of art and imposes it upon nature. In fact it is deduced in *Physics* I, 7, as the necessary condition of any change or becoming. In a compressed way in *Physics* II, 1, and more fully in *Metaphysics* VIII, 2, it is argued that arrangement is insufficient to account for form, which is evident only in the being-at-work of a thing. *Morphē* never means mere shape, but shapeliness, which implies the act of shaping, and *eidos*, after Plato has molded its use, is never the mere look of a thing, but its invisible look, seen only in speech (193a, 31).

fortune *(tuchē)* — Chance which befalls human beings. Luck would ordinarily be a perfectly good synonym, except that Aristotle says that the word has connotations of the divine and of happiness (196b, 5–6; 197b, 4).

genus *(genos)* — A divisible kind or class. It might arise from arbitrary acts of classification, in contrast to the *eidos* or species, the kind that exactly corresponds to the form that makes a thing just what it is. The highest general classes are the so-called categories, the irreducibly many ways of attributing being. *Metaphysics* V, 7, lists eight of these: what something is, of what sort it is, how much it is, to what it is related, what it does, what is done to it, where it is, and when it is. *Categories* IV adds two more: in what position it is, and in what condition.

impasse *(aporia)* — A logical stalemate that seems to make a question unanswerable. In fact, the impasses reveal what the genuine questions are. Zeno's paradoxes are spectacular examples, resolved by Aristotle's definition of motion. In *Metaphysics* III, a collection of impasses in first philosophy, Aristotle writes, "Those who inquire without first being at an impasse are like people who do not know which way they need to walk" (995a, 35–36). The word is often translated as "difficulty" or "perplexity," which are much too weak; it is only the inability to get past an impasse with one's initial presuppositions that forces the revision of a whole way of looking at things.

incidental *(kata sumbebēkos)* — Belonging to or happening to a thing not as a consequence of what it is. The word "accidental" is appropriate to some, but not all, incidental things; it is not accidental that the housebuilder is a flute player, but it is incidental. To any thing, an infinity of incidental attributes belongs, and this opens the door to chance (196b, 23–29).

lead back *(anagein)* — To produce an explanation while leaving the thing explained intact. Aristotle leads back all motion to change of place without reducing all motion to change of place.

material *(hulē)* — That which underlies the form of any particular thing. Unlike what we mean by "matter," material has no properties of its own, but is only a potency straining toward some form (192a, 18–19). Bricks and lumber are material for a house, but have identities only because they are also forms for earth and water. The simplest bodies must have an underlying material that is not bodily (214a, 13–16).

motion *(kinēsis)* — The being-at-work-staying-itself of a potency *as* a potency (*Physics* III, 1–3). Any thing is the being-at-work-staying-itself of a potency as material for that thing, but so long as that potency is at-work-staying-itself as a *potency,* there is motion (1048b, 6–9). Motion is coextensive with, but not synonymous with, change (*metabolē*). It has four irreducible kinds, with respect to thinghood, quality, quantity, and place. The last named is the primary kind of motion but involves the least change, so that the list is in ascending order of motions but descending order of changes.

mover *(kinoun)*
Whatever causes motion in something else. The phrase "efficient cause" is nowhere in Aristotle's writings, and is highly misleading; it implies that the cause of every motion is a push or a pull. In *Physics* VII, 2, it is argued that in one way all motions lead back to pushes or pulls, but this is only a step in a long argument that concludes that every motion depends on a first mover that is motionless (258b, 4–5), and the only kind of external mover that is included among the four kinds of cause in II, 3, is the first origin of motion (194b, 29–30). That there should be incidental, intermediate links by which motions are passed along when things bump explains nothing. That motion should originate in something motionless is only puzzling if one assumes that what is motionless must be inert; the motionless sources of motion to which Aristotle refers are fully at-work, and in their activity there is no motion because their being-at-work is complete at every instant (257b, 9).

nature *(phusis)*
The internal activity that makes anything what it is. The ideas of birth and growth, buried in the Latin origins of our word, are close to the surface of the Greek word, sprouting into all its uses. Nature is evident primarily in living things, but is present in everything non-living as well, since it all participates in the single organized whole of the cosmos (1040b, 5–10). Everything there is comes from nature, since all chance events and products result from the incidental interaction of two or more prior lines of causes, stemming from the goal-seeking activities of natural beings, and all artful making by human beings must borrow its material from natural things.

now *(nun)*
The indivisible limit of a time. The word "moment" is not a suitable translation, because it refers to a *stretch* of a continuous process, nor is the word "instant" appropriate, since the now is relative to a soul that can recognize it. Time arises from the measurement of motion, which can only take place in

a soul that can relate two motions by linking them to a now (223a, 21–26).

number *(arithmos)*
: Any multitude, whether of perceptible things, definite intelligible things, or empty units. The last named, the pure numbers of mathematics, Aristotle calls the numbers *by* which we count (219b, 8), but the word normally refers to the first kind, the numbers which we count, such as the dozen eggs in a carton, a multitude *of* something. The remaining kind of number is alluded to at 206b, 30–33; Plato seems to have taught that higher and lower forms are not related as genus and species but in the same way as a number and its units. That is, the unity of wisdom, courage, temperance, and justice, for example, would not be a common element contained in them all, but the sum of them all as virtue, the eidetic number four. A number in any of its senses is something discrete and countable, and never includes continuous magnitude; it therefore excludes fractions, irrationals, negatives, and all the other things brought under the idea of number when Descartes fused the ideas of multitude and magnitude, or of how-many and how-much, into one. It is thus a paradox, lost on us, when Aristotle says that time is both continuous and a number, but only in resolving that paradox is it possible to see how he understands time.

of-this-kind *(poion)*
: Being of one or another sort is a more direct and immediate feature of things than having a quality (*poiotēs*), a word that Aristotle rarely uses.

original being *(archē)*
: A ruling source; but the word can also mean any beginning or starting point. The usual translations "principle" or "first principle" are rarely adequate, since the word almost never refers to a highest proposition or descriptive rule. The pre-Socratic thinkers posited various original beings as responsible for all being—the one, love and strife, intellect, and so on—as though they were actors in a drama, ancient rulers. Aristotle pokes fun at them in the first chapter of the *Physics*, and argues that our beginnings cannot be *the* beginnings, finally

	deducing that the starting points that should govern an inquiry into things that change are the ideas of form, material, and deprivation. But the whole of the *Physics* becomes a deduction that a ruling original being has been present all along, a motionless first mover.
place *(topos)*	The stable surroundings in which certain kinds of beings can sustain themselves, in which alone they can be at rest and fully active. When displaced, anything strives to regain its appropriate place. This idea of place depends on the prior idea of the cosmos as an organized whole, in which there is no void. The contrary idea of space, as empty, homogeneous, and infinite, Aristotle regards as an abuse of mathematical abstraction: the positing of an extension of body without body.
potency *(dunamis)*	The innate tendency of anything to be at work in ways characteristic of the kind of thing it is; the way of being that belongs to material (1050a, 15). The word has a secondary sense of mere logical possibility, applying to whatever admits of being true (1019b, 32–33), but this is never the way Aristotle uses it. A potency in its proper sense will always emerge into activity when the proper conditions are present and nothing prevents it (1047b, 35–1048a, 16).
primary *(protē)*	First in responsibility. It is translated as "first" when it means first in time.
rest *(ēremia* or *stasis)*	Motionlessness in whatever is naturally capable of motion (202a, 4–5). A natural being at rest is still active. Nothing is inert.
sense perception *(aisthesis)*	Always the reception of organized wholes. Never sensation as meant by Hume or Kant, as the reception of isolated sense-data. The primary object of sense perception is a *this,* a ready-made whole at which one may point.
separate *(chōriston)*	Able to hang together as a whole, intact, on its own. Aristotle never uses the word to mean "separable." Mathematical things are not separate, not because they happen never to be found in isolation, but because they do not compose anything

that could be at work. By the same token, the form *is* separate (1017b, 25–26, and 1029a, 27–30). When the form is remembered or re-constructed in thought as a universal, it is separate only in speech or articulation, but the form as it is in itself, as a being-at-work and a cause of being, is separate simply (1042a, 30–31). In a number of places, such as 193b, 4–5, Aristotle says that form is not separate except in speech, but this is always a first dialectical step, articulating the way form first comes to sight; at 194b, 9–15, he already balances it with the opposite opinion, and points to the inquiry in which the question is resolved.

so much *(poson)*
Not isolated quantity but the muchness or manyness that belongs to something. The former is studied by the mathematician; the latter is present in nature.

thinghood *(ousia)*
The way of being that belongs to anything which has attributes but is not an attribute of anything, which is also separate and a *this* (1028b, 36–37; 1029a, 27–28). Whatever has being in this way is an independent thing. In ordinary speech the word means wealth or inalienable property, the inherited estate that cannot be taken away from one who is born with it. Punning on its connection with the participle of the verb "to be," Plato appropriates the word (as at *Meno* 72B) to mean the very being of something, in respect to which all instances of it are exactly alike. Aristotle elaborates this meaning into a distinction between the thinghood of a thing and the array of attributes—qualities, quantities, relations, places, times, actions, and ways of being acted upon—that can belong to it fleetingly, incidentally, derivatively, and in common with things of other kinds. He concludes that thinghood is not reducible to any sum of attributes (1038b, 23–25; 1038b, 35–1039a, 2). It thus denotes a fullness of being and self-sufficiency, which the Christian thinker Augustine did not believe could be present in a created thing (*City of God* XII, 2); he concluded

that, while *ousia* meant *essentia,* things in the world possess only deficient kinds of being. *Substantia,* the capacity to have predicates, became the standard word in the subsequent Latin tradition for the being of things. A blind persistence in this tradition gave us "substance" as the translation of a word that it was conceived as negating.

this *(tode ti)*
That which comes forth to meet perception as a ready-made, independent whole. A *this* is something that can be pointed at, because it holds together as separate from its surroundings, and need not be constructed or construed out of constituent data, but stands out from a background. The mistranslation "this somewhat" reads the phrase backward, and is flatly ruled out by many passages, such as1038b, 24–25.

underlying thing *(hupokeimenon)*
That in which anything inheres. It can be of various kinds. Change presupposes something that persists. Attributes belong to some whole that is not just their sum. Form works on some material. An independent thing is an underlying thing in the first two ways, but not in the third (1029a, 26–30).

universal *(katholou)*
Any general idea, common property, or one-applied-to-many. It is never separate and can have no causal responsibility, unlike the form, which is a being-at-work present in things, making them what they are (1040b, 27–30; 1041a, 4).

virtue *(aretē)*
Any of the excellences of the human soul, primarily wisdom, courage, moderation, and justice. Though they depend on learning or habituation, Aristotle regards them as belonging to our nature. Without them we are like houses without roofs, not fully what we are (246a, 17–246b, 3).

what it is for something to be *(ti ēn einai)*
What anything keeps on being, in order to be at all. The phrase expands *ti esti*, the generalized answer to the question Socrates asks about anything important: What is it? Aristotle replaces the bare "is" with a progressive form (in the past, but with no temporal sense, since only in the past tense can the progressive aspect be

made unambiguous) plus an infinitive of purpose. The progressive signifies the continuity of being-at-work, while the infinitive signifies the being-something or independence that is thereby achieved. The progressive rules out what is transitory in a thing, and therefore not necessary to it; the infinitive rules out what is partial or universal in a thing, and therefore not sufficient to make it be. The learned word "essence" contains nothing of Aristotle's simplicity or power.

Index

About the Author

Joe Sachs has taught for twenty years at St. John's College, Annapolis, Maryland, where from 1990 to 1992 he held the NEH Chair in Ancient Thought.